MANAGING
RISK
BEST PRACTICES FOR PILOTS

DALE WILSON | GERALD BINNEMA

FOREWORD BY JOHN J. NANCE

AVIATION SUPPLIES & ACADEMICS, INC.
Newcastle, Washington

Managing Risk: Best Practices for Pilots
By Dale Wilson and Gerald Binnema

Aviation Supplies & Academics, Inc.
7005 132nd Place SE
Newcastle, Washington 98059-3153
asa@asa2fly.com | www.asa2fly.com

See the ASA website at **www.asa2fly.com/reader/risk** for the "Reader Resources" page containing additional information and updates relating to this book.

Printed in the United States of America
2016   2015   2014   9   8   7   6   5   4   3   2

Cover photo credit: © Predrag Vuckovic/iStockphoto.com

ASA-RISK
ISBN 978-1-61954-109-2

Library of Congress Cataloging-in-Publication Data:

Wilson, Dale
  Managing risk : best practices for pilots / Dale Wilson, Gerald Binnema.
     pages   cm
  Includes bibliographical references.
  ISBN 978-1-61954-109-2 (pbk.) — ISBN 1-61954-109-2 (pbk.)
  1. Airplanes—Piloting—Safety measures. 2. Airplanes—Piloting—Human factors. 3. Airplanes—Piloting—Handbooks, manuals, etc. 4. Aircraft accidents—Prevention. 5. Risk management. I. Binnema, Gerald II. Title.
  TL710.W55 2013
  629.13028'9—dc23
                        2013042966

This book is dedicated to the memory of Bob Chapman, Ruth Chapman, Teena Daly, Jeffrey Helzer, Al Merali, Danny Penner, Graeme Seath, and Terry Townsend who, like the aviator and author who penned the following words before his life too was tragically cut short in an aircraft accident, *"…slipped the surly bonds of earth…put out [their] hands and touched the face of God."*

Contents

Foreword

BY JOHN J. NANCE

"Every item on this checklist," I recall one of my instructors saying with great serious-ness, "...was written in blood." In other words, someone had to die to get any one of them included.

He was so right, but that truth goes far beyond the authority of checklists and Standard Operating Procedures. In fact, what my IP was pointing to was the invaluable process of avoiding accidents and incidents by absorbing lessons generated by the bad experiences of others.

In a nutshell, that's what this book is about. *Managing Risk: Best Practices for Pilots* cer-tainly weighs in as an extremely well-written and readable work, but what Dale Wilson and Gerald Binnema have produced constitutes far more: It is, in fact, a definitive and invaluable roadmap through a landscape of shattered airplanes and lives in which the telling of each and every incident offers a very real pass to avoiding the same fate.

And given the authors' long and substantial experience training new airmen, investigat-ing accidents and teaching others about flight safety, it is thoroughly authoritative.

Have you ever watched the reaction of a group of pilots—airline, military, or other-wise—when someone casually tosses down a new NTSB report on an air accident or inci-dent? Usually there's a polite scramble to grab it, and the reason is clear: We know that NTSB Reports ("Blue Covers" by the colloquial name) contain a wealth of information about something that went wrong, and we know the information in that publication can, if absorbed and understood, almost certainly guarantee we won't go down the same disas-trous path. While NTSB Blue Covers are *species specific* to a given occurrence, the book you're holding is an easily-absorbed distillation of hundreds of accidents and incidents and a century of evolved knowledge, all for the purpose of imparting experience-generated lessons, methods, and procedures to keep you alive.

And, by the way, knowing those lessons is not just a convenience, it's an obligation for every airman.

Whether you're a newly-minted student pilot or a grizzled veteran with tens of thousands of hours logged, the process of continuously studying and refreshing the knowledge necessary to safely operate air machines is part of the job. This book is an easy and enjoyable way to keep yourself up to speed on the categories of, and the fixes for, the many risks we face. I strongly suggest you read it, tab it, and keep it handy!

John J. Nance
Author of "Blind Trust," "Pandora's Clock," and 17 others.

Acknowledgments

The authors owe a debt of gratitude to our family, friends and colleagues who have supported us in this endeavor. We especially thank our mentors, those pilots—and non-pilots—who encouraged, supported, cajoled and critiqued us as fledgling aviators: Ivan Pettigrew, Jeanette Ritchie, David McKenzie, Joy Carscadden, Harold Faw, and John Horine. Each of you embody what it is to be a positive role model—we couldn't have done it without you.

We thank David Hipschman (and his assistant Colleen Walsh) of the *National Association of Flight Instructors*, and Jackie Spanitz and the team at Aviation Supplies and Academics, Inc., for editing the draft and final versions of this book.

We especially thank John Nance—veteran Air Force pilot and airline captain, prolific author, and a person both authors have admired as one of the world's foremost experts in air safety—for providing valuable feedback and writing the *Foreword* to this work.

Most importantly, we acknowledge and thank God for always being there for us.

About the Authors

Dale Wilson, M.S., is Professor of Aviation at Central Washington University in Ellensburg, Washington, where he teaches courses in flight crew physiology and psychology, threat and error management, aviation safety management, and aviation weather. He holds a Master's degree in Aviation Safety from the University of Central Missouri and a Bachelor's degree in Psychology from Trinity Western University in British Columbia, Canada. He has held several professional pilot certi-  fications including: Airline Transport Pilot, Certified Flight Instructor, Advanced Ground Instructor, and Instrument Ground Instructor from the Federal Aviation Administration; Master Flight and Ground Instructor from the National Association of Flight Instructors; and Airline Transport Pilot License and Class I Flight Instructor from Transport Canada. He has been a pilot for more than 30 years and has logged several thousand hours in single- and multi-engine airplanes in the United States and Canada. He has also served as an Aviation Safety Counselor and later as an FAA Safety Team representative for the Spokane Flight Standards District Office. His primary research interests include visual limitations of flight, pilot decision-making, and VFR flight into instrument meteorological conditions. He has published more than a dozen articles in scholarly journals and professional aviation magazines, and has given numerous safety-related presentations to pilots at conferences and seminars in the U.S. and Canada.

Gerald Binnema, M.A.S., is a consultant in aviation safety, based in British Columbia, Canada. He provides human factors training and assists in developing effective safety management systems for a variety of organizations, including international airports, airlines and helicop-  ter operators. Prior to this he served as an Air Accident Investigator with the Transportation Safety Board of Canada (TSB) and as an Avia- tion Safety Officer with Transport Canada. Mr. Binnema also worked as a flight instructor and in Africa as a missions pilot. He holds a Master's degree in System Safety from Embry Riddle Aeronautical University and a Bachelor's degree from Trinity Western University. He has also studied human factors at Lund University in Sweden. He has been a pilot for more than 30 years and holds a Canadian Airline Transport Pilot License (ATPL); a Class 1 Flight Instructor rating; multi-engine, float, and instrument rat- ings; and a Glider Pilot License and Glider Instructor rating. Mr. Binnema has published several safety articles in aviation publications, including Transport Canada's *Aviation Safety Letter*, and has delivered hundreds of seminars on crew resource management, pilot deci- sion-making, and human factors.

Introduction

This book is for every pilot who wants to avoid an aircraft accident. Whether you are a private pilot who flies a homebuilt aircraft on sunny weekends, an aspiring commercial pilot attending a collegiate aviation degree program, a first officer at your first job at an airline, or a seasoned pilot with thousands of hours under your belt, this book will help equip you with the information you need to successfully manage many of the major risks associated with flight.

The title of this book—*Managing Risk: Best Practices for Pilots*—captures its essence: it documents and describes most of the significant risks associated with flight and, more importantly, provides best-practice countermeasures that you as a pilot can use to avoid or mitigate them.

Flying involves risks. Fortunately, over the history of aviation, most of these risks have been identified—often through tragic losses—and have been managed down to remarkably low levels. Rather than deny the existence of these risks by burying one's head in the sand like the proverbial ostrich, the key is to acquire a thorough knowledge of them and the strategies necessary to identify, eliminate and/or reduce them to acceptable levels. Because "pilot error" is responsible for the majority of aircraft accidents, and because you can't treat an illness without first knowing its cause, this book is as much about discerning the *internal* human limitations of pilot performance as it is about identifying the nature of the *external* threats to safe flight.

After all, there are now very few "new" accidents. If we could apply all the lessons learned from yesterday we could prevent most future accidents, but human beings face a variety of challenges that keep us from consistently carrying out the appropriate action. We are subject to definite physiological and psychological limitations when it comes to piloting an aircraft, so a key component of this book is explaining *why* pilots make the mistakes they do. A more complete knowledge of the external threats to flight safety, coupled with a deeper understanding of how human errors often play out in the cockpit, will help you to successfully manage both.

The book is divided into 10 chapters that cover ten major hazards gathered under four main accident categories: aircraft collisions (runway incursions, midair collisions), adverse weather (aircraft structural icing, VFR flight into IMC, low-level wind shear), physiological hazards (high-altitude flight, night flying, visual illusions, spatial disorientation), and the major threat of controlled flight into terrain.

Using statistics, aviation safety studies and actual aircraft accident examples, each chapter examines the nature of the threat itself, detailing the locations, times or phases of flight where the probability of encountering it is most pronounced. The human aspects that make pilots particularly vulnerable to that specific hazard are also carefully explained. Finally, drawing upon a wealth of expertise and experience provided by government regulatory agencies, the airlines, aviation safety organizations, and the authors' flight experience, each chapter concludes with best-practice strategies that you as a pilot can use to manage the risk.

Each chapter also includes endnote references and a list of resources—such as FAA and NASA online courses—to increase your understanding of that chapter's hazards and the strategies needed to effectively manage them. As a further strategy toward increased comprehension, key terms are indicated in **bold** font throughout the book.

To maintain interest and readability for the average pilot, this book was deliberately not written in the style of an academic textbook. However, as a professional work with a scholarly emphasis, it can be used as a textbook or supplementary text for collegiate aviation safety courses.

Finally, with more than sixty years of combined experience flying, teaching, investigating aircraft accidents, and promoting and writing about aviation safety in both the United States and Canada, the authors wrote this book with pilots from both countries in mind. Perhaps you've heard it said that "England and America are two countries separated by a common language." The same could be said about Canada and the United States. Therefore, for our U.S. readers, you will see the occasional explanation of terms or procedures used in Canada; and for our Canadian readers, you will see U.S. spelling and grammar used throughout. Though primarily written for pilots from these two countries, this book's information and insights will be helpful to pilots and educators worldwide, especially in those regions where the aviation industry is leaving its infancy and beginning to mature. This old adage is universal, applying to aviators everywhere: *"Learn from the mistakes of others. You will not live long enough to make them all yourself."* This book is written to help you accomplish that learning.

A Note About the Accident and Incident Citations in this Book

To better illustrate the nature of the types of hazards pilots face and the errors that they typically make, the authors have cited, within parentheses in the body of the text, numerous aircraft accident reports, primarily from the National Transportation Safety Board (NTSB) and the Transportation Safety Board of Canada (TSB), and incident reports from the National Aeronautics and Space Administration's (NASA) Aviation Safety Reporting System (ASRS). These reports contain a wealth of information about how and why accidents and incidents occur. Those who take the time to track them down and read them will find that most accidents don't happen by accident—they usually result from a variety of human and environmental factors that conspire together and lead to an accident. NTSB and TSB reports can be accessed at *www.NTSB.gov* and *www.bst-TSB.gc.ca*, respectively, and ASRS incident reports can be accessed at *asrs.arc.NASA.gov*. The following are examples of typical accident and incident report citations and how they are "coded."

NTSB/AAR-07/05

The 5th (05) major NTSB aircraft accident report (AAR) issued in 2007 (07).

NTSB-AAR-75-9

The 9th (9) major NTSB aircraft accident report (AAR) issued in 1975 (75). Note: in 1983 the NTSB changed the report number format from hyphens (e.g., NTSB-AAR-82-16) to slash/hyphen/slash (e.g., NTSB/AAR-83/01). Both of these formats are used for major accidents published in Blue Cover Reports, so named because of their blue and white covers.

NTSB Identification No: LAX90LA116

The Los Angeles (LAX) NTSB office filed the accident report, which occurred during the 1990 fiscal year (90). It was a limited aviation accident investigation (LA), the 116th in fiscal year 1990. If the identification number is appended with a final letter, another aircraft was involved in the accident. All NTSB accidents are assigned an accident case number such as this one; however, most major aircraft accidents, especially those involving commercial flights carrying passengers, are identified using the format in the first example above and are published as Blue Cover Reports.

TSB Report No: A04Q0089

A TSB of Canada aviation (A) accident report from the year 2004 (04) in the Quebec (Q) region, which was the 89th accident or incident (0089) in fiscal year 2004.

ASRS Report No: 763177

The report ascension number (ACN) is 763177, which is the 736,177th incident report submitted to the National Aeronautics and Space Administration's (NASA) Aviation Safety Reporting System (ASRS) since the program began in 1976.

The Wrong Place at the Wrong Time

RUNWAY INCURSIONS

It all started with a terrorist bomb. It exploded in the passenger terminal at the Las Palmas Airport in Gran Canaria, a popular tourist destination in the Canary Islands located off the northwest coast of Africa. The airport was closed, and many aircraft were diverted to the Los Rodeos Airport on the nearby island of Tenerife. It didn't take long before this small airport was overwhelmed with aircraft: The apron was crammed with passenger jets, and air traffic controllers (ATC) even had to direct aircraft to park on the only taxiway. Fortunately, within a few hours Las Palmas reopened, and crew members began their preparations for departure.

The air traffic controllers had their hands full, and to make things more difficult, low clouds and fog rolled in reducing their ability to see the runway and making it harder for the pilots to see each other's aircraft.

Flight 4805, a KLM Royal Dutch Airlines Boeing 747, filled mostly with tourists happy to be back on their way to Las Palmas, was cleared to back-taxi (*backtrack* in Canadian terminology) on the runway. A few minutes later Pan American Clipper Flight 1736, another Boeing 747 also filled with tourists glad to be moving again, was cleared to back-taxi on the same runway behind the KLM B-747. The air traffic controller intended for KLM 4805 to taxi to the end of the runway and turn around, while the Pan Am Clipper was to exit the runway and continue to its threshold via the remaining parallel taxiway. Unfortunately, the Clipper crew was having difficulty locating its exit in the dense fog. As a result, the Clipper was still on the runway when KLM 4805 had turned around and aligned for departure at the threshold of Runway 30. Due to a variety of communication problems, the KLM captain thought he had been cleared for takeoff and was certain the Pan Am aircraft was clear of the runway. Unfortunately, he was wrong on both counts. He commenced the takeoff and the aircraft had reached takeoff speed when he first saw the Pan Am jet ahead on the runway. He desperately tried to get his aircraft airborne to clear the other B-747, while the Pan Am crew frantically tried to steer their aircraft off to the side of the runway to avoid a collision, but there simply wasn't enough room—the aircraft collided, killing everyone on the KLM 747 and most of the people aboard the Pan Am 747.[1]

All told, 583 people perished that March afternoon in 1977, making it the deadliest accident in aviation history. Amazingly, it occurred while the aircraft were still on the ground, making it also the worst **runway incursion** accident of all time. Defined several ways since then, the United States and Canada now use the runway incursion definition adopted by the **International Civil Aviation Organization** (ICAO) in 2005. Simply stated, a runway incursion is *the incorrect presence of an aircraft, vehicle, or person on a runway.*

At the time of this writing, the Flight Safety Foundation's Aviation Safety Network has identified the occurrence of more than 65 major runway incursion accidents worldwide that were responsible for the loss of more than 1,250 lives.[2] Seven of these occurred in the United States in one decade alone (the 1990s). Runway incursions don't necessarily result in tragedy; there are literally hundreds of these events each year in the United States and Canada, but fortunately most don't result in accidents. A **surface event** can be classified as a runway incursion even if there is no conflict with a departing or arriving aircraft; the runway environment is intended to be protected, and any time that protection breaks down it results in a potential threat to safety.

The most effective way to reduce the threat of a runway incursion *collision* is to reduce the number of runway incursions themselves. Unfortunately, since the mid-1990s, as airport traffic has increased and airport ground operations have become increasingly more complex, the incidence of runway incursions has increased almost exponentially. Transport Canada (TC) recorded a 145 percent increase in runway incursions in the four years between 1996 and 1999, and the Federal Aviation Administration (FAA) observed a similar trend, albeit not as steep.[3,4]

In response, both organizations implemented significant improvements to the aviation system to help reduce the number of runway incursions (we will look at some of these improvements later in this chapter). However, in spite of their efforts, the number of runway incursions continued to gradually increase in both countries. There were more than 4,200 runway incursions at towered airports in the United States (*controlled* airports in Canada) between 2004 and 2008 with an increase of 39 percent over that five-year period. (The FAA tracks runway incursions only at towered airports, and therefore the statistics do not include the many surface events involving aircraft operating at nontowered airports.)[5] Both the FAA and TC continue to work with airport operators, ground crews, air traffic controllers and pilots to prevent these potentially serious occurrences; however, runway incursions remain a major hazard as evidenced by their presence on the National Transportation Safety Board's (NTSB's) Most Wanted List of safety improvements since it was first created in 1990 until 2013.

What's So Hard About Navigating on the Apron?

Your task is deceptively simple: Navigate from the apron to the active runway so that the "real" flight can begin. After your flight is over and the wheels are back on the ground, it's simply a matter of finding your way to the parking spot. It's this apparent simplicity that leads to a certain degree of complacency about the task. Add to this the fact that the crew might have two or three checklists to complete while navigating across wide open tarmac, often containing confusing signage, and it's easy to see how an error might be made while taxiing.

Environmental Conditions

Most runway incursions occur during the day in visual meteorological conditions (VMC), but statistics indicate that the majority of runway incursion *accidents* occur in conditions of reduced visibility, or at night, or both.[6] In fact, all six fatal runway incursion accidents involving major U.S. air carriers or regional airlines in the 1990s occurred during the hours of darkness (or at dusk) or in fog. It's simply more difficult to navigate on the airport at night or in poor visibility.[7, 8]

Crossing Active Runways

A review of runway incursion accidents and incidents provides dramatic examples of what can go wrong. One of the most common runway incursion scenarios at busy airports involves aircraft that must cross one or more active runways in order to reach their destinations. The following narrative from the captain of a de Havilland DHC-8-30, as reported in the National Aeronautics and Space Administration's (NASA) Aviation Safety Reporting System (ASRS), is a typical example:

> After initiating the takeoff, our aircraft accelerated rapidly due to having a light load for that flight. During the takeoff, a C-182 crossed the runway downfield on what appeared to be Taxiway R. The C-182 was crossing the runway at a rapid rate, and by the time the first officer saw it downfield, it was crossing the center of the runway. We were at or near V_1 at this time, and seeing the aircraft clearing the runway he

elected to continue. He rotated and we were off the ground prior to reaching Runway 5/23. I contacted tower after arriving at our destination and was told that the C-182 was instructed to "taxi and hold short" twice. On both occasions, the pilot read back the hold short instructions and still failed to hold short of our runway (ASRS Report No: 823234).

It was fortunate the Dash 8 was light enough and had the climb capability to clear the Cessna. The Cessna 182 Skylane pilot was twice cleared to "taxi and hold short" of the active runway, and as required by the FAA *Aeronautical Information Manual (AIM)*, he correctly read back his runway hold instructions both times. While taxiing, it's easy to become distracted by tasks such as programming the global positioning system (GPS), completing a checklist, or talking with passengers. You can lose awareness of your position relative to the active runway—especially at an unfamiliar airport—and become aware of the runway only as you cross it. Although hold-short lines are typically marked with yellow lines on the pavement, and red and white holding position signs are located alongside the hold-short point on the taxiway—and in some cases in-pavement lighting is also provided—pilots continue to cross onto active runways, largely due to distraction.

Lining Up and Waiting on the Active Runway

The probability of a ground collision increases when you're cleared to taxi into position on the runway. ATC will often issue such a clearance to provide better spacing between aircraft. Pilots who have been instructed to "**line up and wait**" have lined up with the runway, but all too often have not waited—instead they have initiated their takeoff roll without receiving a clearance. This is often simply a result of habit. Pilots generally enter the runway, line up with the centerline, complete the runway items on the before-takeoff checklist, advance the throttles and depart. This sequence of events becomes so automatic that they sometimes start the takeoff roll even though they haven't been cleared for takeoff.

An almost fatal illustration of this occurred at Quebec/Jean Lesage International Airport when a Cessna 172 was cleared to "taxi to position"[9] on one runway about 16 seconds after an Air Canada Airbus A320 had been cleared for takeoff on another intersecting runway. The Cessna pilot conducted a takeoff departure without a clearance. According to the Transportation Safety Board of Canada (TSB), the controller instructed the pilots of both aircraft several times to abort, but the tower transmitter had been temporarily disabled so the transmissions weren't heard by either of the pilots. The A320 was nearing its rotation speed as it approached the intersection of the two runways, and the Cessna was already in the air climbing through about 200 feet. The captain of the A320 instructed the first officer (FO), who was the pilot flying (PF), to delay rotation until after they had passed the intersection of the two runways. A collision was avoided by only about 200 feet! (TSB Report No: A04Q0089)

Wrong-Runway Departures

Of course, the probability of a collision on the ground increases if pilots taxi to, and depart from, the wrong runway. An early 1990s NASA study of rejected takeoff (RTO) incidents found a surprising number of **wrong-runway takeoffs**—and even *taxiway* takeoffs—conducted by otherwise qualified commercial airline pilots![10] This type of error was more recently committed by the flight crew of Comair Flight 5191 in the early morning hours of August 27, 2006. It didn't result only in the risk of a collision with another aircraft, vehicle or person; it led to something worse.

The crew arrived at dispatch at 5:15 a.m. local time to prepare for a 6:00 a.m. departure from Blue Grass Airport in Lexington, Kentucky. The preflight preparations, including engine start and taxi, were relatively normal, although there were a few minor slip-ups that suggest the crew members may not have been as focused on the task as they optimally could have been. For instance, they initially boarded the wrong airplane and had to be redirected to the aircraft intended for the flight. The radio work and checklists included some minor slips and momentary confusion. According to the NTSB report, the crew also engaged in some non-operational conversations when they should have been focusing on the task of taxiing to the runway. They were cleared to taxi their Bombardier CL-600 Challenger to Runway 22 (the 7,000-foot runway) and cleared to cross Runway 26 during their taxi. The diagram shown in Figure 1-1 illustrates that in order to reach the threshold of Runway 22 from the terminal building, the crew had to taxi past the threshold of Runway 26.

Visibility that morning was about seven miles, but it was dark at the time of takeoff—sunrise wasn't until 7:00 a.m., and the moon was well below the horizon. The taxiway they needed to take (Alpha 7) was a slight left turn, as illustrated in Figure 1-1, but when they encountered the threshold for Runway 26, they turned a full 90 degrees to the left and lined up on the wrong runway. The air traffic controller didn't notice the error and cleared them for takeoff. The result was that Flight 5191 departed on a runway with insufficient takeoff distance (only 3,500 feet long) and hit an earthen berm a few hundred feet past the end of the runway. Everyone on the airplane was killed with the exception of the FO who suffered serious injuries (Report No: NTSB/AAR-07/05).

Wrong-runway takeoffs occur more frequently at airports where two or more runway thresholds are located close together, particularly if one taxiway is used to get to more than one runway, such as at Blue Grass Airport. A short taxi distance from the apron to the runway decreases taxi time and increases workload, thereby increasing the potential for

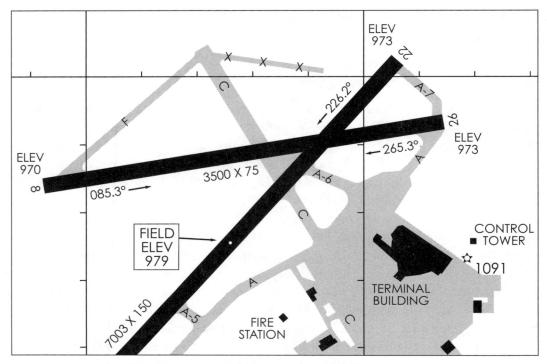

Figure 1-1. Diagram of Blue Grass Airport, Lexington, Kentucky.[11]

confusion. Also, airports with complex layouts can contribute to wrong-runway departures. If an aircraft must make several turns onto different taxiways and cross other runways while en route to the departure runway, the probability of confusion in the cockpit is increased.

Another factor that can lead to a wrong-runway departure is taxiing on a runway. If cleared to taxi via one runway in order to get to another, this increases the possibility that a pilot will depart from the runway on which the aircraft was taxiing. Even airports with a single runway can suffer from wrong-runway departures if there are several intersecting taxiways: A pilot may be cleared for an intersection takeoff and accidentally depart in the wrong direction, 180 degrees from the correct direction.[12]

Controller Error

Pilots aren't the only ones who make mistakes. Air traffic controllers are also human, and their errors, unfortunately, have put pilots and their passengers at risk. Consider the close call that occurred at Vancouver International Airport in British Columbia. Navair Flight

612, a Britten-Norman BN-2 Islander, was in position at the threshold of Runway 08R waiting for takeoff clearance, while Jazz Flight 8191, a de Havilland DHC-8, was also waiting for takeoff clearance on an intersecting taxiway farther along the runway, ahead of Navair 612. The tower controller mistakenly believed the Dash 8 was on the taxiway located at the end of the runway—*behind* the Islander—when he cleared them to taxi into position. He then cleared Navair 612 for takeoff while Jazz 8191 was taxiing to position. Because the taxiway used by the Dash 8 is at an oblique angle to Runway 08R—it's a high-speed exit for Runway 26L—the crew could not see the threshold of the runway or the Islander. However, their clearance specified that they taxi into position behind the Islander, so their visual scan was focused ahead of them as they looked for this aircraft. As they entered the runway they became increasingly uncomfortable because they did not see the other airplane, so they stopped just as Navair 612 rotated for liftoff directly in front of them (TSB Report No: A04P0397).

A tragedy was fortunately averted by the intuition and quick decision-making on the part of the Dash 8 flight crew. To combat what they suspected was a possible error by ATC, they themselves erred on the side of caution and took action. The role of air traffic controllers is highly demanding, is very procedural, and requires high levels of vigilance at all times. Therefore, there will always be occasions when they make mistakes. In fact, runway incursions are classified as pilot deviations (PD), operational (ATC) errors/deviations (OE/D), or vehicle/pedestrian deviations (V/PD). By far, the majority of runway incursions are caused by pilot deviations (63 percent), but 16 percent are attributed to operational errors made by air traffic controllers.[13]

Another error controllers make is clearing an aircraft to land while another aircraft or a vehicle is still on the runway. If the visibility is low because of precipitation, fog or the darkness of night, the landing aircraft may not see the aircraft or vehicle on the ground until it's too late. Such was the case at Los Angeles International Airport in 1991. Night had fallen but visibility was good when SkyWest Flight 5569, a Fairchild Metroliner, was cleared into position on Runway 24L. While its crew was waiting for their takeoff clearance, ATC cleared USAir Flight 1493, a Boeing 737 on final for the same runway, for a visual approach. The tower controller intended to clear SkyWest 5569 for takeoff prior to the USAir's landing but was distracted by several events, including an aircraft transmitting on the wrong frequency and a lost flight-progress strip for another aircraft. As a result, the Metro was still waiting on the runway when the B-737 landed. The subsequent ground collision was the worst runway incursion accident in the United States, claiming the lives of 22 people on the B-737 and all 12 people on the Metro (Report No: NTSB/AAR-91/08).

Nontowered Airports

Runway incursions don't occur only at controlled airports. Reliable statistics on just how many incursions occur at nontowered airports don't exist; however, since the majority of airports do not have a control tower to assist in separating traffic (about 97 percent of airports in the United States are nontowered—almost 20,000 compared with only about 500 towered airports),[14] it stands to reason that the *total* number of incursions is significantly greater at nontowered airports. However, most nontowered airports also have fewer traffic movements, leading to a lower probability of incursions at those individual airports; unfortunately, it's that very fact that can lead to complacency and lax procedures.

It's the pilot's responsibility to follow established procedures at nontowered airports and, if so equipped, maintain a listening watch on the appropriate frequency to identify other aircraft moving on the airport or in the traffic pattern (or *circuit* in Canada). It's obviously important, prior to taking position on the runway, to scan the approach area for other aircraft that may be landing. This is illustrated by an accident that occurred at Fort Collins, Colorado, when two aircraft were in the traffic pattern while a Cessna 152 was waiting to depart. The student pilot in the Cessna observed the first aircraft land but did not see the second aircraft on approach. As the pilot then entered the runway and began her takeoff roll, the second aircraft struck the rotating beacon at the top of the vertical stabilizer and landed directly in front of her. Both aircraft in the pattern were making standard radio calls, but the Cessna 152 pilot did not hear them. The NTSB cited "inadequate visual lookout" on the part of both pilots as the primary cause of this accident (NTSB Identification No: DEN05IA025A).

At a nontowered airport, a mystery often exists about which end of the runway should be active. With a light wind, pilots often choose to land or depart on the runway that is most convenient, but they may forget that other users may choose to use a different runway. In one instance, a motor glider was departing from Runway 11 at Winter Haven Municipal Airport in Florida. The pilot announced his intentions on the common traffic advisory frequency (CTAF), entered the runway and commenced his takeoff roll. The runway has a slight crown at the center, resulting in the opposite end of the runway initially being obstructed from view. As the pilot neared 40 knots he observed another aircraft taking off in the opposite direction on Runway 29, so he veered the glider to the right into the grass, damaging it when the landing gear struck a gopher hole and the wing struck a runway sign. He later asked the pilot of the other aircraft if he had heard his radio calls, but the other pilot said that his aircraft was not equipped with a radio (NTSB Identification No: ERA09CA024).

Radio Communications

Good **radio communication** helps in reducing the probability of runway incursions, but as the preceding two examples illustrate, it's no guarantee. You must always remember that not all aircraft at nontowered airports are radio-equipped. This means that simply listening out for other aircraft is not enough; a careful lookout must also be maintained. In addition, all pilots have at one time or another accidentally pushed a button the wrong way, and thought they had the radio on, the volume set, the correct radio selected on the audio panel, and the radio set to the correct frequency—when they actually didn't. This **finger trouble** has created situations in which pilots were making what they thought were all the right radio calls, but they weren't talking to anybody and nobody was talking to them. Finger trouble can also increase the risk of a midair collision—the topic of our next chapter.

Another issue that creates confusion on the radio—both at towered and nontowered airports—is that of **stepped-on transmissions**. This occurs when two people attempt to transmit on the same frequency at the same time, and the resultant squeal and static on the frequency means that no one hears either transmission. After the FO of the KLM B-747 at Tenerife announced they were taking off, the FO of the Pan American Clipper responded that they were still on the runway, while the tower simultaneously responded by telling the KLM to "… standby for takeoff…." Both transmissions were stepped on, and as a result the KLM crew heard neither transmission and continued the takeoff.

Vehicles, Pedestrians and Animals!

Not all runway incursions are between two aircraft. **Vehicle/pedestrian deviations** (V/PD) cause about 21 percent of runway incursions.[15] Vehicle operators may be driving across a runway at a nontowered airport and not have the aeronautical sense to look out for aircraft on the approach. Vehicle operators at towered airports are required to follow all the same rules as pilots, but they can get into the same trouble as pilots, due to the reasons discussed above. Flight 314, a Pacific Western (PWA) B-737, was on approach to Cranbrook Airport in British Columbia while a snowplow was working to clear snow from the runway. The Flight Service Station (FSS) specialist told the snowplow operator that the expected time of arrival (ETA) of the B-737 was 1:05 p.m., and both expected to hear the aircraft report at a waypoint located about seven minutes from the airport. Unfortunately, the PWA crew didn't make that call and arrived 10 minutes sooner than the ETA. The B-737 landed with the vehicle still on the runway. The flight crew saw the snowplow at the last moment and tried to go around, but one of the thrust reversers didn't retract, and they lost control of the aircraft, killing 42 people in Canada's worst runway incursion accident to date.[16]

It's also necessary to watch out for people and animals on the runway. A pedestrian was killed after he was hit by an Alaska Airlines B-737 conducting a takeoff on Runway 26L at Phoenix Sky Harbor International Airport in Arizona. It was at night, so the pilots didn't see the man, who was not authorized to be on the runway, until they had reached a speed of about 105 knots. The captain immediately veered the airplane to the right and aborted the takeoff. Fortunately, none of the 41 aboard the Boeing jet were killed or injured (NTSB Identification No: LAX90LA116).

Animals frequently get past airport fencing and cross runways at the most inopportune times. Numerous runway incursions have occurred between aircraft and a variety of animals including deer and coyotes, and these collisions can cause major damage to aircraft. Animal control is part of what airport operators do, and controllers at towered airports will likely issue advisories if there are any known wildlife concerns. If you're flying into a nontowered airport, especially if remotely located, it's recommended to check for animals by conducting a low pass prior to landing. This lets you see if any animals are on or near the runway, and it lets the animals know to get out of the way. They dislike collisions as much as you do, and will do what they can to avoid one if they know you're headed their way.

How Things Are Changing

As mentioned earlier, the rapid increase in runway incursions in the mid-1990s prompted the FAA and TC to respond. They set up teams to study the issue, identify challenges and recommend solutions that have resulted in a number of significant changes. Better signage, improved technology, and changes in taxi and communication procedures have all been designed to assist pilots in safely navigating on the airport surface.

Signage and Technology

One of the most obvious improvements is improved surface markings and signage at all of the busiest airports in North America. Markings on the pavement have been clarified and refreshed to make them harder to miss when you're crossing onto a runway. Signage has been improved and standardized according to ICAO guidelines. Stop bars, consisting of a series of red in-pavement lights installed across the entire width of the taxiway and above-ground lights located on the side of the taxiway, have been installed at certain high-density traffic thresholds and intersections. Pilots should never cross a stop bar when illuminated, even if an ATC clearance has been given to proceed onto or across the runway (FAA *AIM*).

Several technological aids have been installed at a number of busy airports. **Airport surface detection equipment** (ASDE) uses special ground radar systems to track all surface-based movements and warns controllers of potential conflicts or incursions. A more precise system—Model X (ASDE-X)—takes advantage of multiple sources of information including radar as well as **automatic dependent surveillance-broadcast** (ADS-B) systems to accurately pinpoint the position of all ground-based movements. ADS-B will be explained more fully in the next chapter as a tool to help prevent midair collisions. The FAA is also looking at lower-cost options for some of the less busy airports.

Airport Safety Assessments

Early research into this threat found that some airports are more conducive to incursions than others. In the United States, for example, Los Angeles, Chicago O'Hare and Lambert–St. Louis International Airports ranked first, second and third, respectively, for the highest number of runway incursions between 2001 and 2004. However, during that same period, Chicago O'Hare and Hartsfield–Jackson Atlanta International Airports ranked first in the number of most serious **Category A incursions** (where a collision was only narrowly avoided).[17] Many major airports have been carefully assessed, and some have been found to have a disproportionate number of runway incursions due to the way the taxiways and runways are configured. These assessments have resulted in changes in procedures, signage and even taxiways in order to reduce the risk of runway incursions. Locations on airport-movement areas with a history or potential risk of collisions or runway incursions, and where heightened attention by pilots/drivers is necessary, have been identified as airport surface **hot spots**. These areas are now depicted on U.S. airport diagrams and Canadian aerodrome charts.

ATC Communication

In order to reduce the risk of runway incursions, changes were made in the way ATC provides clearances and in the response they require. No longer are you allowed to cross a runway without specific ATC authorization, and no longer will ATC issue such an unrestricted clearance. Controllers are now required to issue explicit instructions to cross or hold short of *each* runway—active, inactive or even closed. Instructions to cross multiple runways will not be issued, and an aircraft or vehicle must have crossed the previous runway before another runway-crossing clearance is issued. The clearance will usually go something like, "Taxi Charlie and Bravo to Runway 08, hold short of Runway 12." You must read back the hold-short clearance, and ATC will demand that you do. They're not trying to be difficult—they just want to make sure that you know you must stop before crossing Runway 12.

Another recent change is the substitution of the phrase "cleared to position and hold" with "line up and wait." In all other parts of the world, when ATC wants pilots to taxi onto the active runway and wait for their takeoff clearance, they use the phrase "line up and wait." Only the U.S. and Canada were using the "position and hold" phrase, and this caused the potential for confusion on the part of pilots from other parts of the world who were not accustomed to using that terminology.

Avoiding a Runway Incursion

Fortunately, you can implement several strategies to minimize the probability of a collision while maneuvering on the airport. Taking certain steps before flight and incorporating **best practices** while taxiing will help. So will practicing good communication skills and using technology effectively—both will increase your awareness of others' locations on the apron and their knowledge of yours, thereby decreasing your odds of running into each other.

Prepare Before Your Wheels Leave the Chocks

Being well-prepared beforehand will prevent preoccupation with last-minute duties, enabling you to better focus on the primary task of taxiing your aircraft safely to its destination on the airport. Study the airport diagram (aerodrome chart) during your preflight planning and look for some of the traps that lead to runway incursions or wrong-runway departures: taxiways that take you by more than one runway threshold, very short taxi routes that tend to increase your workload, and routes that result in crossing active runways or cause you to taxi on a runway. Beware of two different runway thresholds that are almost at the same location. One of the older classic airport designs included three runways shaped like a triangle, so that runway thresholds were very close to each other. These airports often produce a disproportionate number of runway incursions due to wrong-runway departures.

Keep the airport diagram in front of you while taxiing in order to maintain your positional awareness. Airport diagrams for most airports are available for free from the FAA and Nav Canada websites (see links in the Helpful Resources section at end of this chapter). Finally, look for any hot spot information depicted on the airport diagram, aerodrome chart or taxi chart. These areas are circled and a brief textual description is provided. Hot spot markings are published by the FAA in the U.S. *Airport/Facility Directory* (*A/FD*) and by Nav Canada, the Canadian air traffic control agency, in the *Canada Airport Charts* and the *Canada Air Pilot*.

Complete as many items on the checklist as you can before beginning to taxi, especially if you're flying as the single pilot or you're unfamiliar with the airport—and don't rush your taxi. If you get a taxi clearance with a very short route, rather than attempting to rush the checklist—which will only divert your attention away from navigating to the runway—stop and complete the checklist prior to beginning your taxi. If you get a complex taxi clearance and are unfamiliar with the route, make sure to write it down. You can also ask for "**progressive taxi instructions**." These are precise step-by-step taxi directions given to you in stages as you progress across the airport. You might think you're creating extra unnecessary work for the ground controller, but he or she would rather have you follow progressive instructions than get lost on the airport.

Practice Good Taxi Techniques

Don't taxi too fast. When the Boeing 747 was first introduced to the market more than 40 years ago in 1970, it was reported that pilots were taxiing too fast, putting too much strain on the landing gear while turning. Being the first wide-bodied aircraft with an **eye-to-wheel height** (height of pilot's eyes above the apron) almost two times higher than the previous narrow-bodied craft, the **optic flow** of the apron's surface area moving past them in their visual field created the illusion that they were taxiing at the same speed they used to. The same phenomenon occurs at night or in poor visibility, even if you pilot a small airplane with an eye-to-wheel height of only a few feet. Since an aircraft's airspeed indicator doesn't give any indication of the actual speed while taxiing, the almost total absence of optic flow in your peripheral vision in these conditions creates the illusion of taxiing too slow, resulting in higher taxi speeds. Since it's easy to miss a sign or a taxiway marking in these conditions, avoid taxiing too fast, otherwise you may inadvertently end up on an active runway.

Keep a sharp lookout while taxiing. Prior to crossing a runway, make sure you have been cleared—never cross a hold line without *explicit* ATC instructions; if in doubt, ask. Even if you are cleared across, ATC or another pilot could be making an error, so always check the approaches and thresholds of both ends of the runway for landing or departing aircraft. If you're flying in a **crewed flight deck** (two or more flight crew members required), good **crew resource management** (CRM) includes calling out the runway crossing by saying "all clear my side" or words to that effect. (Basic CRM concepts will be further explored in Chapter 2 on evading a midair collision and in Chapter 10 on avoiding **controlled flight into terrain**.) It was precisely this habit that saved the lives of the people on the Dash 8 at Vancouver International Airport in the example referenced earlier: even though the airplane was cleared to enter the runway, the crew took the time to visually check for another aircraft, and this practice saved them from a collision.

Remember to check your compass as you line up with the runway. The **compass check** ensures your heading indicator is set properly (or slaving properly) and you are lined up on the correct runway. On one occasion, Gerald Binnema, one of this book's authors, was departing at night from an airport where the thresholds of Runway 01 and 07 were accessed from the same taxiway. He was distracted by a passenger and accidentally lined up on Runway 07 when he had been cleared for Runway 01. His habitual use of the compass check identified the problem before it grew into a possible catastrophe. This check is a basic habit that you should practice as part of your departure routine. However, it falls prey to complacency because our heading indicator is almost always set correctly, and we tend to always believe we're departing from the correct runway. It's not that we become lazy; it's just that it's difficult for us to maintain vigilance when we see no apparent problem. In fact, you may fly your entire professional career, and the compass check may identify a problem only once or twice. But when it does, it could save your life. Consistently conducting the check is vital and requires discipline. It should also be verbalized during takeoffs in a crewed flight deck and is typically part of the required **call-outs** in an operator's **standard operating procedures** (SOP). Unfortunately, there was no evidence on the cockpit voice recorder (CVR) that the crew of **Comair Flight 5191** had performed such a compass/heading check to confirm their airplane was on the assigned runway; perhaps if they had, the tragedy at Blue Grass Airport on that morning in Lexington would have been averted.

If you're cleared to line up and wait on a runway, get in the habit of feeling slightly vulnerable out there. Have a good look for aircraft on approach before you enter the runway. If you're cleared to line up from a taxiway intersection, look for other aircraft waiting for takeoff farther down the runway behind you. The FAA *AIM* recommends that pilots turn on their landing lights during takeoff, either after the takeoff clearance has been received or when beginning the takeoff roll. However, at the discretion of the pilot-in-command, the *AIM* also recommends turning on all external illumination, including landing lights, when holding in position on any runway. Activating all available aircraft lighting such as navigation/position lights, anti-collision lights, strobes, beacons and recognition lights significantly increases your aircraft's conspicuity and enables other pilots to see your aircraft in position on the runway. This might have prevented the tragic collision between the Fairchild Metroliner and the Boeing 737 in Los Angeles mentioned earlier. As you recall, the SkyWest Metro was cleared into position while the USAir B-737 was on the approach to the same runway. According to the NTSB, the Metro's navigation/position lights and red anti-collision beacon located on top of the vertical stabilizer were the only lights illuminated on the airplane at the time of the collision. Had the USAir crew been able to see the Metroliner earlier, they possibly could have avoided the collision.

Communicate Effectively

Good **communication** habits save lives. They not only help reduce the risk of ground collisions, but as you will see throughout this book, they also reduce the risk of many of the hazards encountered in aviation. Effective communication skills that reduce the threat of a runway incursion include operating your communication radio properly, using your call sign, following recommended communication procedures, maintaining a listening watch on the appropriate frequency at all times, and maintaining a sterile cockpit environment.

Earlier in this chapter we mentioned how finger trouble when operating radios can lead to a runway incursion: You can select the wrong frequency or even the wrong radio, or the volume can be so low that you transmit but are unable to hear any transmissions received. It takes discipline to maintain the habits necessary to avoid making these careless mistakes. Every time you establish communication with a new ATC facility, even when you switch from ground to tower frequency, do a quick check. You can check that the radio is on, the volume set and the correct radio selected with one action: turning on the squelch function. If you hear static when you turn it on, you know that all of the above are set correctly. After that, all you need to do is verify that the proper frequency is selected, and you're ready to transmit.

However, before you press the push-to-talk button, always listen out on the frequency before transmitting in order to make sure you're not interrupting a conversation. Besides causing a great deal of unnecessary duplication of effort, stepped-on transmissions can lead to a disaster, as seen in the Tenerife accident. If you hear an ATC transmission that is stepped on, and you don't think ATC is aware of it, quickly transmit a terse statement saying simply, "Two on at once."

Always use your call sign when transmitting. Use your *full* call sign (tail number) on initial contact and abbreviate it only when ATC does after communication has been established. If there are other aircraft with similar-sounding call signs, then keep using your full call sign to avoid confusion. According to the FAA *AIM*, you should use the phrase "verify clearance for (your complete call sign)" if you aren't sure the clearance was meant for you.

Follow the recommended procedures outlined in the FAA (or TC) *AIM* regarding communicating at an airport and consult the appropriate advisory circulars (see Helpful Resources at the end of this chapter) regarding recommended communication procedures at nontowered airports and at airports with towers that are closed. Had the crew of PWA Flight 314 reported at the beacon on final approach to the runway—which was the normal recommended and expected call—the snowplow operator would have exited the runway and the tragedy at Cranbrook Airport on that cold snowy day in February of 1978 would most likely never have occurred.

Don't automatically relax your guard and "tune out" once you've established communication with ATC or received your clearance. If you pay attention to the other calls being made, you can construct a mental picture of the traffic moving about the airport. This helps reduce the chance of an incursion, particularly if ATC erroneously gives you a clearance that could lead to a conflict with another aircraft.

Finally, there's a reason why it's illegal in many jurisdictions to operate hand-held cellular devices while driving an automobile: distractions cause accidents. The same is true for flying. In its report on Eastern Air Lines Flight 212, a DC-9 that collided with terrain in the morning fog three miles short of the runway at Charlotte/Douglas International Airport, the NTSB found that conversations not pertinent to the operation of the aircraft contributed to the accident. The crew talked about politics, used cars and the location of a local theme park tower (Report No: NTSB-AAR-75-9).

This accident—which claimed the lives of 71 people—and many others like it prompted the FAA in 1981 to publish what is commonly known as the **sterile cockpit rule**. Required for Part 135 and 121 operations, this regulation (14 CFR §§135.100 and 121.542) prohibits crew members from engaging in nonessential activities (including extraneous conversations) that could distract them from completing the essential duties required for the safe operation of their aircraft during the critical phases of flight. A **critical phase of flight** is one in which the level of risk is elevated—during periods of high workload when crew members are more likely to make mistakes and when the aircraft is operated closer to the ground. These include all ground operations—taxi, takeoff and landing—and all other operations (i.e., climb, descent, approach) below 10,000 feet, except cruise flight.

Unfortunately, since its adoption, noncompliance with the sterile cockpit rule has been implicated in a number of major accidents. The NTSB concluded that nonpertinent conversation during taxi resulted in the flight crew's loss of positional awareness, leading Comair Flight 5191 to depart on the wrong runway at Blue Grass Airport in Lexington. Had they followed the sterile cockpit rule while taxiing their Challenger to the runway, perhaps they would have made it to the correct runway. There are enough interruptions (e.g., from passengers, flight attendants, dispatch, ATC) to distract you from your performance, so why self-distract yourself and your crew with conversations not related to the flight? Even if you aren't an airline pilot, adhering to this rule will go a long way in helping you avoid an accident during taxi, approach, landing and other critical phases of flight.

Sometimes it takes tact, diplomacy and **assertiveness** to inform passengers or others of your need to focus on flying the aircraft; unless it is important to the safety of flight, they should refrain from interrupting that task. Tell your passengers during your preflight briefing that there will be times when you will not be able to engage in conversation with them because your attention must be focused on your flying. The sterile cockpit rule is mentioned in several other chapters in this book because complying with it is an important means of managing distraction during critical times.

Use Technology to Assist You

Some electronic aids can help increase your positional awareness. Besides providing you with all the relevant information you need to know about the airport, the electronic flight bag (EFB) displays a moving map of the airport with your aircraft's position depicted on it. Designed to replace most paper documents found in a pilot's flight bag, EFBs also contain everything you need for flight including such items as the aircraft flight manual (AFM), the company's operations manual, SOPs, weight and balance, and aircraft performance calculation programs. You may not have an electronic flight bag readily available, but if your aircraft is equipped with a GPS map display, you can zoom in to see your location on the airport.

A fatal accident at the Quincy Municipal Airport in Illinois provides serious food for thought regarding many of the risk factors we've mentioned in this chapter. A Great Lakes Aviation Beechcraft 1900, doing business as United Express 5925, was approaching the nontowered airport with the crew planning to land on Runway 13. Two other aircraft were departing the airport: a Beechcraft King Air A90 and a Piper Cherokee. With a wind from 060 degrees at about 10 knots, both were planning to use Runway 04. Runways 04 and 13 cross each other, intersecting about 2,000 feet from their respective thresholds. The captain of the arriving B-1900 made appropriate calls and twice asked conflicting traffic to report, but she received no response. The pilots of the two departing airplanes made calls announcing their plans to taxi to Runway 04, and the King Air A90 announced it was taxiing into position on Runway 04. Unfortunately, there is no indication that the crew of the King Air heard any broadcasts from the other aircraft. The NTSB concluded that the King Air occupants—the pilot and pilot-rated passenger—likely did not properly configure the radio receiver switches to the common traffic advisory frequency (CTAF), or they were preoccupied, distracted or inattentive.

The crew of Flight 5925 observed the King Air A90 taxi into position on the intersecting runway and asked them if they were departing or were going to hold until their B-1900 had landed. The King Air pilot did not respond, but for some reason the Cherokee pilot in the run-up area behind the King Air did, indicating that he was holding behind the King Air. Unfortunately, in the middle of that broadcast, a **ground proximity warning system** (GPWS) announcement came on in the flight deck of the B-1900, and the portion of the broadcast that the crew heard left them with the impression that it was the King Air that was holding, not the Cherokee. The King Air began its takeoff roll about 13 seconds before Flight 5925 touched down. It does not appear that the pilots in the King Air saw the B-1900, even though the Cherokee pilot later testified that the B-1900 was quite visible with its landing lights on (the sun had just set and the weather was VFR with good visibility of 12 miles). The B-1900 used maximum braking but was unable to stop before the intersection, and the two aircraft collided. Both aircraft caught fire. It's likely that all 10 passengers and two crew members in the Beech 1900, as well as the two pilots in the King Air A90, survived the initial collision but died from smoke inhalation in the post-crash fire.

The NTSB concluded that the probable cause of the accident was the "failure of the pilots in the King Air A90 to effectively monitor the common traffic advisory radio frequency or to properly scan for traffic, resulting in their commencing a takeoff roll when the Beech 1900C was landing on an intersecting runway" (Report No: NTSB-AAR-97/04).

This tragic accident underscores the significance of many of the risk factors and countermeasures we've highlighted. For example, you should exercise caution and pay extra attention when operating at airports with intersecting runways and complex layouts. You need to also use effective communication skills and adhere to recommended communication procedures, especially at airports without operating towers. If a transmission is garbled, cutoff or stepped on, you should immediately seek clarification. Had the crew of the Beech 1900C sought clarification of the message they thought was from the King Air, and had the pilots in the King Air been maintaining a good listening watch on the frequency or had they made a good visual scan of the approach end of intersecting runways during their takeoff run, it's likely this accident would not have occurred.

Perhaps the point most clearly underscored by this tragedy is the relative simplicity of runway incursions. A moment's lack of attention can cause a catastrophic collision on the airport, but a little care and attention can also prevent one. Pilots often think flight only consists of the time between wheels-up and wheels-down, and forget to use the strategies needed to keep from being in the wrong place at the wrong time during those crucial moments when navigating to and from the runway.

Helpful Resources

The FAA's Office of Runway Safety website contains a variety of excellent resources, including training materials, videos and actual runway incursion animations designed to help you avoid a collision while operating on the ground. (www.FAA.gov/airports/runway_safety) Many of its runway safety videos can also be seen on YouTube (www.YouTube.com) including:
- *Face to Face, Eye to Eye*
- *Explicit Runway Crossing*
- *Line Up and Wait*

Line Up and Wait, an online course developed by the FAA Safety Team (FAASTeam), presents best practices during taxi operations including coverage of the new rules requiring ATC authorization before crossing runways and the procedural change from "taxi into position and hold" to "line up and wait." (www.FAASafety.gov/gslac/ALC/course_catalog.aspx)

The AOPA Air Safety Institute has two free online interactive courses that can increase your understanding of the runway incursion threat and improve your communication skills. (www.AOPA.org/asf/online_courses)
- *Runway Safety*
- *Say It Right: Mastering Radio Communication*

AOPA Air Safety Institute Safety Advisors can assist you in understanding and reducing the hazard of runway incursions. (www.AOPA.org/asf/publications/advisors.html)
- *Collision Avoidance: Strategies and Tactics*
- *Operations at Nontowered Airports*
- *Operations at Towered Airports*

The International Civil Aviation Organization's (ICAO) Runway Safety Site contains several resources, including toolkits and manuals designed to help pilots, ATC and airline operators combat the threat of runway incursions. (www2.ICAO.int/en/RunwaySafety)

Advisory circulars published by the FAA and Transport Canada (TC) offer guidance on how to safely navigate on the airport.

For the U.S.: www.FAA.gov/regulations_policies/advisory_circulars/

For Canada: www.TC.GC.ca/eng/civilaviation/standards/commerce-circulars-menu-284.htm

- *Single Pilot, Flight School Procedures During Taxi Operations* (FAA AC 91-73)
- *Flightcrew Procedures During Taxi Operations* (FAA AC 120-74)
- *Traffic Advisory Practices at Airports without Operating Control Towers* (FAA AC 90-42)
- *Recommended Standard Traffic Patterns and Practices for Aeronautical Operations at Airports without Operating Control Towers* (FAA AC 90-66)
- *Situational Awareness During Taxi & Take-Off* (TC AC 722/723/724/725/604)

Airport diagrams and hot spot information for many airports are available for free from the FAA's Runway Safety website (www.FAA.gov/airports/runway_safety) and from Nav Canada's website (www.NavCanada.ca).

Notes

1. Spanish Commission on Accidents, *KLM Boeing 747 and Pan Am Boeing 747 Collision at Los Rodeos Airport (Tenerife) Spain on 27 March 1977* (Civil, Madrid, Spain: Subsecretaría de Aviación Civil, November 16, 1978).

2. Flight Safety Foundation, Aviation Safety Network Database, *Collision—Aircraft: On Ground (Runway Incursion)*; *Collision—Object: Person, Animal*; *Collision—Object: Vehicle (on Runway)*. Available online at Aviation-Safety.net/database/events/event.php?code=CO.

3. Transport Canada, *National Civil Aviation Safety Committee Sub-Committee on Runway Incursions: Final Report* (Ottawa, ON: TC, September 14, 2000).

4. Federal Aviation Administration, *Despite Significant Management Focus, Further Actions Are Needed to Reduce Runway Incursions*, Report No. AV-2001-066 (Washington, DC: FAA, June 26, 2001).

5. U.S. Department of Transportation, Office of Inspector General, *DOT's Fiscal Year 2009 Top Management Challenges*, Report No: PT-2009-005 (Washington, DC: U.S. DOT, November 17, 2008). Available online at www.OIG.dot.gov/library-item/4474.

6. International Civil Aviation Organization, *Manual on the Prevention of Runway Incursions*, Doc 9870, AN/463 (Montreal: ICAO, 2007): 2-2.

7. Flight Safety Foundation, *Collision—Aircraft*; *Collision—Object*.

8. Dale R. Wilson, "Darkness Increases Risks of Flight," *Human Factors and Aviation Medicine* 46 (November–December 1999): 1–8. Available online at FlightSafety.org/archives-and-resources/publications/human-factors-aviation-medicine.

9. The phrase used to be "taxi to position and hold," but has since changed to "line up and wait," in both Canada and the United States. "Taxi to position," a shortened version of "taxi to position and hold," was also acceptable ATC phraseology in Canada.

10. Roy W. Chamberlin, "Rejected Takeoffs: Causes, Problems, and Consequences," *Flight Safety Digest* 12 (January 1993): 1–7.

11. Federal Aviation Administration, *Wrong Runway Departures* (Washington, DC: FAA, July 2007).

12. Ibid.

13. Federal Aviation Administration, *Annual Runway Safety Report 2010* (Washington, DC: FAA, 2010).

14. AOPA Air Safety Institute, *Safety Advisor: Operations at Nontowered Airports*, Operations & Proficiency No. 3 (Frederick, MD: AOPA ASI, January 2007).

15. Ibid.

16. Flight Safety Foundation Aviation Safety Network, *Accident Description* (Pacific Western Airlines, B-737, Cranbrook Airport, BC, Canada, February 11, 1978). Available online at Aviation-Safety.net/database/record.php?id=19780211-0.

17. Air Line Pilots Association International, *White Paper: Runway Incursions, A Call for Action* (Washington, DC: ALPA, March 2007).

The Big Sky Is Not So Big

2

They were considered the most advanced passenger transport aircraft of their time. Built by two of the world's giants in the aerospace industry, Lockheed Corporation's Constellation (dubbed the "Connie") and the Douglas Aircraft Company's DC-7—both large piston-driven propeller airplanes—set records for speed, distance, altitude and passenger-seating capacity. On the morning of June 30, 1956, both models were represented at the Los Angeles International Airport, each operated by a different airline. Flight 2, a Trans World Airlines (TWA) L-1049 Super Constellation (a "stretch" version of the Connie) carrying six crew members and 64 passengers, departed at 9:01 a.m. bound for Kansas City. Flight 718, a United Air Lines (UAL) DC-7 carrying five crew members and 53 passengers, departed three minutes later for Chicago. Initial routing placed the DC-7 south of the Connie, but since Chicago is north of Kansas City their paths would eventually cross somewhere near the Grand Canyon in Arizona. UAL Flight 718 flew at a cruising altitude of 21,000 feet while TWA Flight 2 was to fly at 19,000 feet. However, the crew of Flight 2 encountered some convective activity and requested an altitude change to 21,000 feet. This was initially denied by ATC because of the potential conflict with the DC-7 flying at that same altitude. However, after they requested a clearance to fly "1,000 feet on top," and just before leaving controlled airspace and positive control by ATC, they were advised of the DC-7's position and their request was granted.

ATC and company personnel became concerned when neither aircraft provided their scheduled position reports. After an unsuccessful communications search, a missing aircraft alert was issued, followed by the initiation of search and rescue procedures. Their worst fears were confirmed after the pilot of a small airplane discovered the wreckage of both airplanes located about a mile from each other at the eastern end of Grand Canyon

National Park. There were no survivors: All 128 people on both airplanes perished, making this, at the time, the worst civilian aircraft accident in history.[1]

The Civil Aeronautics Board (CAB) determined that the aircraft had collided at 21,000 feet. It concluded that neither the crew nor the air traffic controllers had done anything particularly wrong; the two aircraft were flying direct routes in uncontrolled airspace in visual weather conditions and were responsible for providing their own separation from other aircraft. The CAB investigators determined that the cause of this accident was that the pilots of both airplanes simply failed to see each other in time to avoid a collision.

This tragic event proved to be a major turning point in aviation safety in the United States. It was the catalyst for dramatic improvements to the country's National Airspace System (NAS). During the few years that followed, significant changes were made: The new Federal Aviation Agency was created (it later became the Federal Aviation Administration in 1967), taking over most of the duties of the Civil Aeronautics Administration (CAA), and it was given complete jurisdiction over the NAS; funding was provided for many new long-range radar facilities and tower radar units and almost double the number of very high frequency omnidirectional radio ranges (VORs); the use of secondary surveillance radar was developed enabling ATC to positively identify transponder-equipped aircraft; all continental airspace above 24,000 feet (and later above 18,000 feet) was positively controlled; better separation was provided between VFR and IFR traffic; and new regulations designed to reduce **midair collisions** (MACs) were enacted.

Due to the sparsity of evidence, including the lack of credible eyewitnesses to this accident, investigators could only speculate as to why the pilots, in day VFR conditions, were unable to see each other. Therefore, this accident also prompted new regulations requiring all large commercial airplanes (those with a maximum certificated takeoff weight, or MCTOW, of more than 12,500 pounds) to be equipped with **flight data recorders** (FDRs) by 1958, and **cockpit voice recorders** (CVRs) by 1964. Other improvements included the establishment of terminal control areas (TCAs, now called **Class B airspace**) at most of the nation's busiest airports, with positive ATC control of all aircraft and the requirement for operable Mode C (altitude encoding) transponders for participating aircraft. By the late 1970s, a conflict alert system that warned air traffic controllers of a collision threat using visual indications and aural alarms was also implemented at all air route traffic control centers (ARTCCs).

These measures were effective in reducing the incidence of MACs. However, worldwide air traffic continued to grow at a phenomenal pace—more than threefold in just the 10 years between 1959 and 1968[2]—and, unfortunately, so did the number of MACs. In 1968 they had peaked to almost 40 in the United States. In fact, between the time of the Grand Canyon accident and another historic collision that occurred over the skies of Southern

California more than 20 years later, the United States lost more than 1,300 lives in about 550 midair collisions.[3]

It was a warm sunny morning in late September when it happened. Pacific Southwest Airlines (PSA) Flight 182, a Boeing 727 carrying 128 passengers and seven crew members, was on an extended downwind for Runway 27 at Lindbergh Field (San Diego International Airport). A Cessna 172, with an instructor and student aboard, was climbing away from the airport on a vector of 070 degrees for an eventual return to Lindbergh for a third practice instrument landing system (ILS) approach. San Diego Approach Control provided several advisories to the crew of PSA Flight 182 regarding the location of the Cessna ahead. Just before the approach controller handed them off to Lindbergh Tower, the first officer told the captain, "Got 'em," and the captain informed the controller that they had the "traffic in sight." San Diego Approach then informed the Cessna: "Traffic at six o'clock, two miles, eastbound; a PSA jet inbound to Lindbergh, out of three thousand two hundred, has you in sight." Less than a minute later a conflict alert warning went off in the San Diego Approach Control facility indicating that the predicted flight paths of Flight 182 and the Cessna would conflict. The approach controller responded by issuing a second advisory to the Cessna: "Traffic in your vicinity, a PSA jet has you in sight, he's descending for Lindbergh." Unfortunately, the advisory was too late—the two aircraft had already collided. All told, 144 people—all 137 people aboard both airplanes and seven people on the ground—perished in what was at the time the worst collision in U.S. history (Report No: NTSB-AAR-79-5).

Many of the improvements made since the Grand Canyon accident were in play on that fateful day in 1978: Both aircraft were talking to, and were radar-identified by, ATC; both were equipped with operating Mode C altitude encoding transponders; and the San Diego Approach facility was equipped with a conflict alert system designed to warn the controllers—and by extension, the pilots—of a collision threat. However, the Cessna was operating VFR, and the crew of the B-727 had accepted a visual approach with the instruction by ATC to "maintain visual separation." In other words, the pilots of both aircraft were responsible for *visually* maintaining their own separation. Both the tower and approach controllers saw the close proximity of the two airplanes on their radar screens, but based on the information that was communicated to them by the flight crew, they both believed the pilots of Flight 182 had the Cessna in sight and were maintaining visual separation. In fact, after advising Flight 182 of the Cessna traffic ahead, the tower controller heard this response: "He's passing off to our right." What neither controller knew was that the crew members had lost sight of the Cessna, and after discussing among themselves where it might have gone, they made the assumption that it must have passed off to their right. The NTSB determined that the probable cause of the accident was "the failure of the flightcrew of Flight 182 to comply with the provisions of a maintain-visual-separation clearance, including the requirement to inform the controller when they no longer had the other aircraft in sight." Contributing to the accident were ATC procedures that allowed the controllers to use "visual separation procedures to separate two aircraft on potentially conflicting tracks when the capability was available to provide either lateral or vertical radar separation to either aircraft" (Report No: NTSB-AAR-79-5).

Fortunately, this was the last fatal MAC in the United States involving a major *U.S. air carrier*. The last MAC in U.S. airspace involving *any* major scheduled commercial airline occurred eight years later in 1986—also over the sunny skies of Southern California—when an AeroMexico DC-9 collided with a Piper Archer that had inadvertently entered the Los Angeles TCA without a clearance. The collision killed 15 people on the ground in the town of Cerritos and all 68 people aboard the two airplanes (Report No: NTSB/AAR-87-07).

After the PSA accident, a TCA with its requirement for positive ATC separation of all aircraft—both VFR and IFR—was established over San Diego airspace; operable Mode C transponders were required for all aircraft operating within 30 nautical miles (NM) of a TCA's primary airport; and, even though it had been contemplated before the Grand Canyon collision,[4] the use of airborne **traffic-alert and collision avoidance systems** (TCAS)—known internationally as airborne collision avoidance systems (ACAS)—was mandated in 1993 for all airliners with more than 30 passenger seats operating in U.S. airspace. The rule was further expanded in 1996 to include commercial turbine-powered airplanes with 10 to 30 passenger seats, and because of the significant growth in air freight traffic, in 2003 expanded to cargo airplanes weighing more than 33,000 pounds MCTOW.

As a last line of defense for preventing in-flight collisions, TCAS is independent of any onboard navigation equipment or ground-based ATC radar. It interrogates ATC transponders of nearby aircraft and displays potential collision threats within a volume of airspace around TCAS-equipped aircraft. TCAS II provides two types of advisories: Traffic advisories (TAs) display visual information to alert pilots of other transponder-equipped aircraft in their vicinity to assist them in visually acquiring the intruding aircraft and to prepare them for a possible resolution advisory (RA). If the threat aircraft comes within 25 seconds from the closest point of approach—a definite collision threat—TCAS issues an RA in the form of voice commands and visual guidance (usually red lights on the vertical speed indicator to indicate vertical speeds to be avoided, and green "fly-to" lights to indicate the desired vertical speed to be flown) to help the crew in avoiding a collision. TAs will be generated only for intruder aircraft equipped with operating transponders (Mode A, C, or S), and RAs will be issued only when the transponder of the intruding aircraft is providing altitude information (Mode C or S). If the intruder aircraft is also equipped with TCAS II, the respective computers will coordinate opposite-direction resolution maneuvers for each aircraft.[5]

Though not a completely foolproof system, the fact that there have been no MACs involving TCAS-equipped Part 121 scheduled airline flights in the United States since the 1993 mandate is testimony to its effectiveness. Its use has since been made compulsory in many countries (including Europe in 2000[6] and Canada in 2007[7]), making the occurrence of MACs among major commercial air carriers a relatively rare event. In fact, there have been only three major MACs involving commercial airlines since the mid-1990s—a B-747 and an Ilyushin near New Delhi, India, in 1996; a B-757 and Tupolev over Überlingen, Germany, in 2002; and a B-737 and an Embraer Legacy in Brazil in 2006. However, these three accidents were responsible for the loss of 574 lives, which makes in-flight collisions still one of the major causes of worldwide commercial airline fatalities. In fact, the MAC near New Delhi, with the loss of 349 lives, ranks as the world's third worst aviation accident to date.

The reduced incidence among airlines doesn't mean that in-flight collisions no longer pose a threat to safe flight operations—they still do, and when they do they are often deadly. A search of the various accident databases reveals that between the time of the U.S. TCAS mandate and this writing, there were more than 200 MACs in the United States and more than 30 in Canada. The vast majority of these involved general aviation (GA) aircraft, while a few involved military, air taxi or other types of nonscheduled commercial operations. Also, even though MACs account for less than one percent of all GA accidents, they frequently result in fatalities. A study of 329 in-flight collisions that occurred in the United States between 1983 and 2000 found that compared to an overall GA fatal accident rate of less than 20 percent, 56 percent of in-flight collisions resulted in fatalities with the loss of 570 lives.[8]

Even though the number of MACs are relatively low, the number of "near misses" (a misnomer—"near hits" would be more accurate) aren't. According to the FAA *AIM*, a **near midair collision** (NMAC) occurs when two aircraft come within less than 500 feet of each other or when, in the opinion of either pilot, they come close enough to create a collision hazard. During the 10 years between 2000 and 2009, an average of 145 NMACs per year were reported in the United States with 39 per year involving Part 121 scheduled air carriers.[9] Most safety experts agree that since these reports are voluntary, the majority of NMACs are not reported, especially among the GA population.

The number of aircraft and number of hours flown over the next 20 years are expected to double for the world's commercial jet fleet[10] and modestly expand for the U.S. GA fleet.[11] As air traffic continues to grow in airspace that doesn't, the probability of experiencing an in-flight collision will only increase. Therefore you as a pilot will have to work even harder—and more importantly, smarter—to ensure that the aircraft you fly doesn't attempt to occupy the same space as another.

Midair Collision Risk Factors

The MAC hazard has been studied by accident investigators, aviation safety specialists, and scientists almost from the time of the first aircraft collision. The data reveal certain situations that increase the likelihood of experiencing a MAC. Being aware of these risk factors is the first of many steps you should take in developing strategies to minimize your chances of experiencing an in-flight collision.

High-Density Traffic Areas

Some pilots reason that since the sky is so big and the number and size of aircraft are so small, the chance of experiencing a MAC must be infinitesimally low. So they reason, "Why should I expend so much energy on preventing one?" The so-called "big sky" theory, however, falls apart when you realize that, just as with automobile traffic, the likelihood of a collision significantly increases in areas where aircraft are concentrated. Accident statistics confirm this: Navigation aids such as VORs are magnets that draw aircraft into midair collisions, but to an even greater degree, so do airports.

An early NTSB study found that almost 75 percent of in-flight collisions during 1969 and 1970 took place within five miles of an airport. The study also revealed that almost 65 percent occurred in the traffic pattern with three-quarters of those occurring during the final approach and landing phases of flight.[12] Fast-forward more than three decades: An Aircraft Owners and Pilots Association (AOPA) Air Safety Institute study found that most MACs occur within five miles of an airport and below 3,000 feet above ground level (AGL), and about half occur in the traffic pattern area with most of those on the final approach and landing where a faster aircraft usually overtakes a slower one.[13,14] It appears that not much has changed: The closer you get to the runway, the greater your chances of colliding with another aircraft.

The AOPA study also found that about 78 percent of traffic pattern collisions take place at nontowered airports. This corresponds with statistics elsewhere. For example, over a 10-year period in Canada, the TSB discovered that two-thirds of in-flight collisions (that didn't involve formation flights) occurred in the vicinity of uncontrolled airports.[15] Also, Australia's equivalent of the NTSB—the Australian Transport Safety Bureau (ATSB)— found that 78 percent of MACs since 1961 in that country occurred in or near the traffic pattern with more than one-third occurring on final approach or the base-to-final turn.[16]

This is not to say that there are no in-flight collisions in the big sky of the enroute environment—there are—but the probability increases the closer you get to where aircraft tend to gather.

Day VFR Weather

Since most MACs occur during the times when, and at the locations where, air traffic is at its greatest—especially GA traffic—it shouldn't be surprising that most occur during the warmer months of spring and summer, on weekends, in good VFR weather conditions, and during the hours of daylight.[17] The latter suggests it is likely easier to see other aircraft at night than during the day—provided their position and navigation lights are on! Also, with the exception of the visual portions of the flight, IFR flights benefit from the traffic separation services of ATC.

Recreational and Training Flights

Statistics demonstrate that if you are a pilot who flies primarily for personal or recreational purposes, you are at an increased risk of experiencing a MAC. For example, a study of 79 MACs that occurred in the United States between 1994 and 1999 found that 93 percent of the accident aircraft were being flown for personal use.[18] Another notable statistic is that even though flight instruction accounted for less than 10 percent of total civilian flight hour activity, the NTSB found that 29 percent of MACs occurring between 1996 and 2005 involved instructional flights.[19] The AOPA Air Safety Institute also found that 16 percent of dual and 20 percent of solo fatal instructional accidents were in-flight collisions, and that even though fewer than 10 percent of U.S. pilots are flight instructors, more than one-third of all MACs had an instructor onboard at least one of the aircraft.[20] This was a problem before the San Diego accident (where there was an instructor aboard the Cessna 172 and a student wearing a view-limiting device, i.e., a hood) and is still a problem today. Teaching in the cockpit involves special challenges, but maintaining separation from other aircraft is by far one of the greatest.

Formation Flights

It's obvious that the probability of two aircraft colliding in flight increases the closer they fly to each other. It should come as no surprise, then, that even though **formation flying** accounts for a small fraction of total flying activity, it is responsible for about 17 percent of all MACs.[21] The margin of error when flying in close—or even loose—formation is so small, that any mistake, miscalculation or unexpected turbulence could lead to catastrophe. Formation flying requires considerable training and experience, yet these types of collisions have claimed the lives of professional air show pilots and skilled military pilots. What is also tragic is they have claimed the lives of amateurs, many of whom had no formal training in formation flying techniques.

See and Avoid: How Good Are We at It?

According to the U.S. right-of-way rules published in 14 CFR §91.113(b), "When weather conditions permit, regardless of whether an operation is conducted under instrument flight rules or visual flight rules, vigilance shall be maintained by each person operating an aircraft so as to **see and avoid** other aircraft." (emphasis added) A cursory glance at NMAC reports indicates that many in-flight collisions were averted because one or more pilots saw the other in time to take evasive action; however, the most common cause of MACs cited by accident investigators is still the failure of pilots to "see and avoid" other aircraft. For example, of the 329 MACs in the study cited earlier in this chapter, 88 percent of the pilots failed to see the other aircraft in time to avoid the collision.[22]

The logical assumption is the pilots involved simply weren't looking; if they had been, they would have *seen and avoided* the other aircraft. However, from what we've learned since the Grand Canyon accident, even though keeping a sharp lookout for other traffic is critical in avoiding an in-flight collision, in many cases it is not enough. For example, the Massachusetts Institute of Technology conducted a study to determine the ability of pilots to detect other aircraft on a near-collision course. Twenty-four volunteers—half of whom held commercial pilot certificates—each flew a 45-minute cross-country in a Beech Bonanza with an observer/safety pilot. Participants were told the objective of the study was to measure pilot workload and, among other things, were instructed to immediately notify the observer whenever they saw another aircraft, whether it appeared to conflict with their flight path or not. Once established at cruising altitude, and unbeknownst to the pilots beforehand, a Cessna 401 flew by their airplane within a few tenths of a mile horizontally and at 500 feet above or below their altitude—once from head-on, once from directly behind and once at an angle from the side. Even though the pilots knew one of their tasks was to point out any air traffic, only 56 percent of the target aircraft were detected.[23]

Safety experts have learned a great deal about the nature of in-flight collisions over the years and have concluded that the "see and avoid" concept has significant limitations, especially as the sole means of avoiding a collision. It turns out that a variety of environmental, aircraft and physiological factors decrease our ability to visually detect other aircraft in flight.

Environmental Factors

Environmental factors such as lack of contrast, poor visibility and glare reduce a pilot's ability to detect and avoid a collision with another aircraft.

Lack of Contrast

Reduced contrast diminishes the level of aircraft conspicuity; a bright aircraft in front of a bright sky (or dark aircraft in front of a dark one) or an aircraft that is positioned in front of a cluttered background (e.g., buildings) reduces the level of contrast and is more difficult to detect. For example, investigators suspected the PSA crew likely didn't see the Cessna, in part because it "blended in with the multicolored hues of the residential area beneath" (Report No: NTSB-AAR-79-5). Also, since many collisions occur on exceptionally clear days, experts suspect that these conditions don't actually reduce the risk but may actually increase it! One theory is that an aircraft seen on the horizon in extremely clear conditions lies in front of a darker background providing less contrast, while one seen on the horizon during normal visibility conditions, in which a layer of brighter haze is usually present on the horizon, provides more contrast.[24]

Reduced Visibility

Poor visibility makes it difficult to see other aircraft and provides less room to maneuver should a conflicting target be detected. Also, the phenomenon of **aerial** (or atmoshperic) **perspective**, where distant objects appear darker and more blurred in conditions of reduced visibility, causes targets to appear farther away than they actually are. This could cause you to unnecessarily delay your response to a collision threat.

Glare

Glare from sunlight or from its reflection off objects interferes with the ability of the human eye to see. The ability to see when flying directly into the sun or above an overcast sky can be particularly difficult. Restricted vision from glare was cited as a factor in the collision of an Atlantic Southeast Airlines Embraer EMB-120 Brasilia and a Civil Air Patrol Cessna 172 over Gadsden, Alabama. The Cessna pilot was flying west into the setting sun and took no evasive action prior to impact; the Brasilia was able to land safely, but both occupants of the Cessna died (NTSB Identification No: ATL90FA095A). Glare from the setting sun also hampered the ability of the pilot of a Beechcraft B35 Bonanza to see a Cessna 172 traveling in the opposite direction. The two collided over the north shore of the St. Lawrence River near Mascouche, Quebec, killing both occupants of the Cessna.[25]

Aircraft Factors

Limits in our ability to see and avoid also arise from features of the aircraft itself. For example, some cockpits provide a better view out the window than others, and a crazed, pitted or dirty windscreen, no matter how large, hampers our ability to effectively see outside.

A Bonanza pilot reported a windscreen post and hardware attached to a vent window obstructed his view of another aircraft; luckily, he saw it just in time, but missed it by only an estimated 10 feet! (ASRS Report No: 763177)

High Wing Versus Low Wing

Perhaps you remember seeing the pictures of what was later dubbed the "Cess-per." A low-wing Piper PA-28 Cadet descended onto a high-wing Cessna 152 while on final approach to Plant City, Florida, and the two airplanes got stuck together! Fortunately, the pilot of the Cessna managed to successfully land the two aircraft without injury to the occupants. This was the quintessential example of several collisions that have occurred because of the inherent blind spots created when a low-wing airplane flies above a high-wing—only this one had a happy ending.

Design Eye Position

There is an ideal seat adjustment that provides not only the best access to the flight controls, but also an optimum viewing angle for both cockpit instrumentation and the outside environment. Some aircraft manufacturers specify where the pilot's eyes need to be to obtain this optimal vision. It is called the **design eye position** (DEP)—also called the design eye reference point (DERP)—and if your eyes are not positioned at this point in space, your ability to adequately see outside the cockpit will be hindered. For example, your over-the-nose field of view below the horizon will be diminished if your seat is positioned too low, bringing your eyes below the DEP.

Physiological Limitations

Physiological factors can also severely limit our ability to see and avoid other aircraft. Students of the human eye know it to be one of the most fascinating and complex organs in the body; yet even with two of them, there are limitations in our ability to accurately "see" what's really outside of the cockpit. We will revisit this important topic in later chapters when discussing night flying and visual illusions, but for now it's important to understand some of the reasons why *what you see is not always what you get*. The effect of any one of the following factors is enough to impair your ability to detect traffic, but if multiple factors are present at the same time, their cumulative effect is even greater.

Accommodation Lag

To focus clearly on objects that vary in distance, the shape of your eye's lens must change. Muscles change the shape of the lens to focus light waves on the retina. This **accommodation** process takes time, so when attempting to focus outside after looking inside at the instrument panel, there is a time lag before proper focusing takes place. Even though this may only be a fraction of a second to two seconds in duration, the results could be significant if the time available to take evasive action is minimal.

Empty-Field Myopia

As its name suggests, **empty-field myopia** (sometimes called empty-sky myopia, night myopia, or dark focus) is a nearsightedness that occurs when looking outside into an empty visual field. This can happen when flying at high altitudes (where terrain and cloud features are absent), flying in conditions of reduced visibility (such as in mist or haze) or flying during the hours of darkness. The absence of distinctive objects in an empty visual field of view causes the eye to focus at its resting state which, depending on the individual observer, ranges from only two feet to two yards away. The result is that objects farther away than this distance are blurred or even imperceptible.[26] In these conditions, you might be under the false impression that you're conducting an adequate visual lookout when in fact you're not. This looking-without-seeing is further aggravated by the fact that this phenomenon makes distant objects appear smaller and farther away than they actually are—certainly not something that is conducive to avoiding a collision. Windscreen posts and frames, and dirty windscreens, also tend to pull eyes' focal distance inward.

Blind Spot

Visual blind spots are created not only by the structural features of the aircraft, but also because of the structure of the human eye. The optic nerve delivers the raw visual signals from our eyes into the brain; in the small area where the nerve is connected to the retina we are blind. This blind spot is located approximately 18 degrees from the center of the retina and is usually unnoticeable since one eye fills in what the other can't see. However, if vision from one eye is blocked (by a windscreen post, for example), a target in the periphery of vision may be completely undetectable. In fact, a jumbo jet will be undetectable as close as 1.5 miles away if its image falls on the blind spot. See Figure 2-1 to discover the existence of the blind spot in your own eyes.

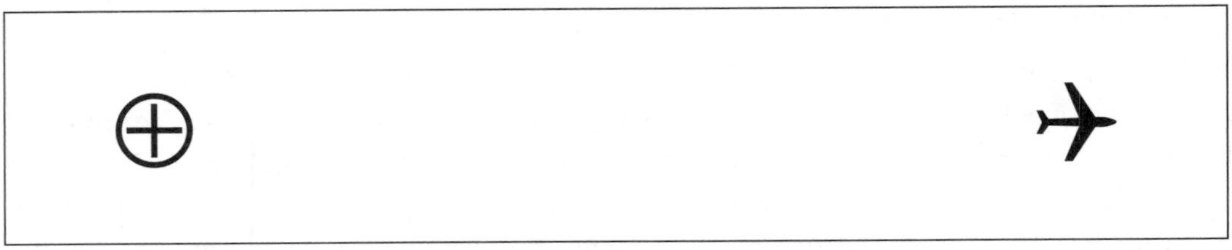

Figure 2-1. Hold the page close to your face with your left eye closed. Stare at the center of the circle with your right eye while slowly moving the page away from you; the airplane in your peripheral vision will disappear when its image falls on the blind spot.

Central Vision Blindness at Night

When quickly traveling from a well-lit environment into darkness (into a darkened theater, for example), the ability to see is initially severely impaired, but improves with time. That's because within the retina of each eye are two distinct types of sensory systems: one responsible for **day vision (cones)** and another for **night vision (rods)**. As light levels drop after sunset, the cone receptors responsible for day vision gradually stop functioning while the rod receptors continue to increase their sensitivity until full **dark adaptation** is complete. This usually takes about 30 minutes but can take considerably longer in the case of previous exposure to extreme lighting conditions. This is another reason why it is advisable to wear sunglasses during a day at the beach if a night flight follows.

Rods make up the bulk of the periphery of the retina but are completely absent from its center. Cones are concentrated in a small portion of the center of the retina called the fovea. This is the part of vision that provides the highest resolution, but the area of coverage is surprisingly small. Since there are no rods in the fovea, and the foveal cones are shut down during night vision, staring directly at a dimly lit object at night (a star, for example) will make it seem to disappear. Therefore, during pure night vision there are two blind spots in each eye.

Peripheral Vision Limitations

Visual acuity is the ability to discriminate fine detail, but it is only good (in daylight) up to a maximum of one to two degrees from the center of the fovea. (This roughly corresponds to the area of a quarter held at arm's length.) Visual acuity drops sharply the farther an image is from the fovea, until at about 15–20 degrees of arc the maximum possible visual acuity attained is about 20/200. The farther out an object is located from the foveal field of view, the more difficult it is to perceive. In fact, the FAA *AIM* states that an aircraft at a distance of seven miles that appears in sharp focus when a person is looking directly at it with foveal vision would need to be as close as seven-tenths of a mile away in order to be recognizable when not being looked at directly. Therefore, you should realize that you may not always see an aircraft that is in your peripheral vision.

Peripheral vision is important in maintaining spatial orientation, but is really only effective in detecting moving targets, not ones that appear stationary on the retina. In a recent collision involving a Piper PA-25 Pawnee and a Cirrus SR20 near Boulder, Colorado, the NTSB determined that the probable cause was the failure of both pilots to "see and avoid" each other, in part because there was no perceived visual relative motion of each airplane by the pilots; both aircraft remained stationary in each other's windscreen right up to the time of the accident (NTSB Identification No: CEN10FA115C).

Relative Motion of Target

If an automobile remains at a fixed location through the view of your window and increases in relative size as it enters the on-ramp to the highway, a collision is unavoidable unless one of you moves. Similarly, if the relative bearing of another aircraft remains constant (i.e., relative motion equals zero) and it's increasing in size—as was the case in the Piper and Cirrus accident above—then an in-flight collision will occur. Unfortunately, the target that is most difficult to detect in our peripheral vision (i.e., zero relative motion) is also the one we are most likely to hit.

Relative Size of Target

Although the relative size of an approaching aircraft doubles whenever its distance from the viewer is reduced by half, its apparent size (i.e., visual angle on the retina) remains relatively small until it is quite close. Its size then increases exponentially and appears to blossom (see Figure 2-2). This **blossom effect** has been the experience of many who have survived a MAC or NMAC. Despite maintaining what they thought was an adequate visual lookout, pilots have reported that they were surprised to find an airplane appear out of nowhere "filling up" their windscreen. One pilot got so close he could actually see the other pilot; he said he "wore a blue shirt and what looked like aviator Ray-Bans!" (ASRS Report No: 787228).

Avoidance Response

A pilot's ability to respond to the threat of a MAC is first impeded by the difficulty in detecting other aircraft and then further delayed by the relatively long time needed to take effective evasive action. According to the FAA Advisory Circular (AC) 90-48, *Pilots' Role in Collision Avoidance*, to recognize a target as an aircraft, become aware of the collision course, make a decision to take evasive action, and implement that action takes a minimum of 12.5 seconds. This reaction time is based on military data, so it would likely be even longer for less maneuverable GA or air transport aircraft. It certainly isn't enough time if

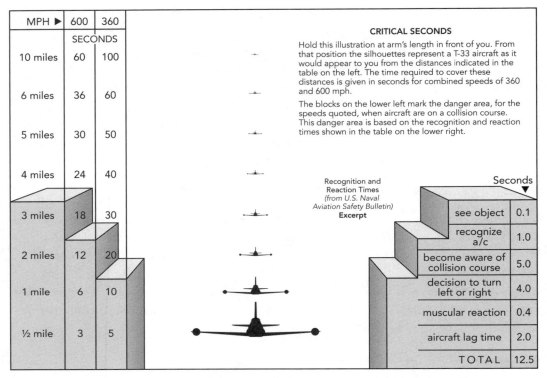

MPH ▶	600	360
	SECONDS	
10 miles	60	100
6 miles	36	60
5 miles	30	50
4 miles	24	40
3 miles	18	30
2 miles	12	20
1 mile	6	10
½ mile	3	5

CRITICAL SECONDS

Hold this illustration at arm's length in front of you. From that position the silhouettes represent a T-33 aircraft as it would appear to you from the distances indicated in the table on the left. The time required to cover these distances is given in seconds for combined speeds of 360 and 600 mph.

The blocks on the lower left mark the danger area, for the speeds quoted, when aircraft are on a collision course. This danger area is based on the recognition and reaction times shown in the table on the lower right.

Recognition and Reaction Times
(from U.S. Naval Aviation Safety Bulletin)
Excerpt

	Seconds ▼
see object	0.1
recognize a/c	1.0
become aware of collision course	5.0
decision to turn left or right	4.0
muscular reaction	0.4
aircraft lag time	2.0
TOTAL	12.5

Figure 2-2. Relative size of approaching target and pilot reaction times for avoidance.[27]

aircraft closing speeds are high, or if the other aircraft is only recognized after it blossoms in the windscreen (see Figure 2-2).

Many of the weaknesses of the see-and-avoid concept were at play in a recent collision of a Piper PA-32 Lance with a Eurocopter AS350 helicopter over the Hudson River in New Jersey. According to the accident report, the Lance was located behind and outside the helicopter pilot's field of view for the last 32 seconds of the flight; the helicopter remained below the horizon against a cluttered background of buildings and appeared relatively small and stationary in the airplane's windscreen, and the Piper pilot likely didn't see the Eurocopter until it blossomed in the windscreen, which was, by then, too late. All occupants of both aircraft—three in the airplane and six in the helicopter—were killed in the accident. In its probable cause statement, the NTSB cited only two reasons for this accident: failure of ATC to adequately conduct its duties, and the "inherent limitations of the see-and-avoid concept, which made it difficult for the airplane pilot to see the helicopter until the final seconds before the collision" (Report No: NTSB/AAR-10/05).

Finally, you shouldn't think that just because there are serious limitations to the see-and-avoid concept that maintaining a vigilant lookout isn't important in avoiding a collision—it certainly is. But as we have seen, sometimes it isn't enough. Therefore, in addition to being vigilant in using proper visual scanning techniques, it is also important to use other strategies that have proven to be helpful in minimizing your chances of experiencing an up-close-and-personal encounter with another aircraft.

Avoiding a MAC

Effective strategies to reduce the probability of experiencing a MAC include developing habits to help you improve your visual scanning performance; using effective verbal communication techniques to help others know your location and taking advantage of auditory information to assist you in knowing theirs; and following regulations and recommended procedures designed to minimize the chance of experiencing an in-flight collision.

See and Be Seen

Even though the availability of technology has increased our ability to avoid a MAC, as we have already mentioned, regulations still acknowledge the reality that whether you are flying VFR, or in visual meteorological conditions (VMC) when flying IFR, you are required to maintain vigilance and have the responsibility to *see and avoid* other aircraft.

Use an Effective Visual Scanning Technique

The systematic use of an effective **visual scanning procedure** improves your ability to visually spot other aircraft. Our eyes cannot adequately focus when using wide sweeping eye movements; they need to momentarily stop and focus. An AOPA *Safety Advisor* recommends you scan the entire outside visual field by dividing the sky into 10 to 15 degree "blocks" and then pause at each one, looking approximately 10 degrees above and below your flight path for at least one to two seconds to allow for proper refocusing and the utilization of foveal vision. It also recommends starting from one side and working your way across to the other (side-to-side scan), or looking directly ahead of you and scanning all the way to one side before returning to the center and scanning to the other (front-to-side scan).[28]

AOPA's Air Safety Institute also found that contrary to popular belief, only a small percentage of MACs involve head-on collisions; most—*82 percent*—involve faster aircraft overtaking a slower one from behind.[29] Even though it is important to scan the entire sky, studies show that most pilots don't—their gaze tends to default directly ahead with many glances to only a few degrees to either side.[30] Therefore, it is crucial to scan your entire field of view and move your head as far left and right as you can during the scan. Also, if you see your aircraft's shadow on the ground or on top of a cloud layer, keep an eye out for other shadows that may be converging on yours.

To see what may be located in a blind spot, you should also periodically perform gentle turns and/or momentarily level off during prolonged climbs or descents. Be sure to clear the area (by raising a high-wing or lowering a low-wing) before every turn and perform clearing turns before conducting any type of training or other maneuvers.

If you are flying in empty-field conditions (high altitude, haze, or night), you should refocus your eyes on distant objects (e.g., the ground or even a wingtip) about every two minutes since it takes approximately that long for your eyes to return to their resting state. For dimly-lit objects at night, you might need to use off-centered vision for easier identification; looking slightly to the side of an image will project it on the peripheral rods, thus making identification easier. However, developing this off-centered viewing takes some effort since we are strongly conditioned to look directly at objects.

You can improve your ability to see in bright sunlight and when encountering glare by wearing good quality sunglasses, but be aware that some make it more difficult to see reflected light off other aircraft—this is one of many reasons why polarized lenses are not recommended for aviators. Choosing the right type of lens is not always as easy as it sounds, so consult an ophthalmologist who is knowledgeable about the advantages and disadvantages of a particular type for the flight environment.

Before departure, clean your windscreens and windows of all dirt and bugs and remove any obstructions to vision—such as clipboards and charts—and make sure you adjust your seat to the correct DEP. Many transport category airplanes are equipped with a fixed eye level indicator located at eyelevel above the center glareshield. It's usually just a matter of placing your head in the normal position while adjusting your seat until the appropriate-colored ball is superimposed directly over the other. Consult the AFM if your aircraft is not fitted with such a device; failing that, contact the manufacturer for guidance. If you want to find the ideal seat position on your own, the Transport Canada (TC) *AIM* suggests you adjust your seat so it is comfortable, allows unrestricted use of all flight controls through their full range of motion, provides an unobstructed view of all flight instruments and warning lights, and allows optimal out-of-the-cockpit visibility without being restricted by the nose of the aircraft.

Don't Get Buried in the Cockpit

The odds of spotting other aircraft increase when you spend more time looking outside; therefore, it's important to avoid unnecessary preoccupation with activities inside the cockpit, especially in situations where the risks are greatest (e.g., during takeoff and approach, in the traffic pattern, near VORs, etc.). The FAA *AIM* recommends you spend at least 65 to 75 percent of your time looking outside scanning for traffic; however, experimental studies indicate that most pilots do just the opposite—they spend close to that percentage of time looking *inside* at the instrument panel instead.[31]

You can avoid getting your head buried in the cockpit through effective cockpit management strategies. Review performance figures, charts, navigation logs, airport diagrams, frequencies and checklists during your preflight planning, and properly organize access to these materials in the cockpit so the time needed to look at them while flying will be minimized.

Preoccupation inside the cockpit is no doubt the reason why the risk for a MAC doubles for instructional flights;[32] these involve at least two pilots engaged in the activity of teaching and learning and by necessity involve a greater degree of focused attention inside the cockpit. Checking out in a new or more advanced aircraft also involves a higher degree of preoccupation with in-cockpit activities. The bottom line: Even if it's more challenging to maintain a vigilant lookout when participating in an instructional or checkout flight, you must make the extra effort if you're going to avoid a collision.

Recruit as Many Eyes as You Can

Over the past three decades, airline pilots have been encouraged to practice good crew resource management (CRM) skills to minimize the number of errors made in the cockpit. A classic definition of CRM is *the effective use of all available resources—people, equipment, information—to achieve safe and efficient flight operations.* These concepts have recently been adapted to single-pilot operations in the form of **single-pilot resource management** (SRM). The FAA's definition of SRM is very similar to CRM: "SRM is the art and science of managing all the resources (both onboard the aircraft and from outside sources) available to a single pilot (prior and during flight) to ensure that the successful outcome of the flight is never in doubt."[33]

Both definitions emphasize the importance of utilizing the resources available to you. Your passengers are one of those valuable resources and can assist in your search for potential traffic conflicts, so why not enlist their help? Without unduly alarming them, ask them to keep an "eye out" and let you know if they see any traffic.

You are at least eight times more likely to see another aircraft if you are alerted to its presence.[34] That's why you should ask ATC for its help in searching for potential collision threats—even if flying VFR. In the United States the FAA encourages VFR pilots to request radar **traffic advisories** (also called flight following) whenever possible to help alert them to the presence of potentially conflicting traffic. These advisories will usually give the location (in terms of the 12-hour analog clock) and distance (in NM) of another aircraft, its type and altitude (if known), and the direction it is traveling. An example of such an advisory is: "Traffic, one o'clock, five miles, westbound, six thousand."

Consult the *AIM* for more information about this service and how to obtain it, but keep in mind the following points: First, do *not* assume that ATC will advise you of all traffic—the service is on a workload-permitting basis. The controller may be preoccupied with higher-priority duties, some aircraft may not be talking to ATC or some may not even appear on the controller's radar screen. Therefore, as the FAA *AIM* reiterates, if you are operating in VMC, radar service does not relieve you of the responsibility to see and avoid other traffic—whether flying VFR or IFR. So don't let down your guard just because ATC is also looking for traffic.

Second, radar advisories are based on course (track) not heading. If a north wind causes one aircraft to fly a heading of 070 degrees to maintain an easterly track of 090 degrees and causes opposite traffic to fly a heading 290 degrees to maintain a westerly track of 270 degrees, ATC will tell both that the other is located at their "twelve o'clock" when in fact one will be at *eleven o'clock* and the other at *one o'clock*.

Finally, even if you sight an aircraft, keep looking for others nearby; there could be traffic that the controller failed to inform you of, or you may have identified the wrong aircraft. Though ruled out by the majority report, evidence of a possible third unknown aircraft led one NTSB investigator to speculate, in his dissenting opinion, that the crew members of PSA Flight 182 may have mistaken it for the Cessna, which could explain why they thought it had passed and was no longer a factor. This is still a problem today, as the pilot of an Airbus A320 who identified the wrong aircraft recently discovered: the jet narrowly missed colliding with a twin Beechcraft by 200 feet because of it (ASRS Report No: 858282).

Be Seen by Others

If you can't spot others, help them to spot you. Required for night flying, aircraft position lights can also increase your aircraft's conspicuity during the day. Both the FAA and TC also encourage the use of landing lights—*day or night*—in conditions of reduced visibility and for all takeoffs and landings. The FAA *AIM* recommends using them below 10,000 feet, especially within 10 miles of any airport, and near designated VFR checkpoints and waypoints; the TC *AIM* recommends their use below 2,000 feet within terminal areas and aerodrome traffic zones. The use of supplementary strobe lights is also an effective way to attract others' attention. While position lights are hard to see in the daytime, and landing lights are only visible when looking at the front of an aircraft, strobe lights are bright and flashy, and are usually visible for 360 degrees. Consult the *AIM* for further recommendations on the use of aircraft lighting systems.

Be Seen Electronically

Many MACs could have been prevented if both aircraft's transponders were on. Fifty people lost their lives when a Hughes Air West DC-9 collided with an F-4 Phantom jet in Southern California; five people died when a Cessna Citation collided with a Cessna 172 in Atlanta's Class B airspace; and 154 people were killed in the world's worst midair since 1996 when a Gol Airlines B-737 collided with an Embraer ERJ Legacy 600 business jet over the Amazon jungle in Brazil. All of these—and several more—occurred in part because a transponder was either inoperable or off. Not only is it good safety practice, but the regulations now require pilots of all aircraft equipped with a transponder to operate it—including Mode C if installed—on the appropriate VFR code or ATC-assigned code at all times while in controlled airspace. Consult the *AIM* for the recommended procedures to follow.

Hear and Be Heard

The whereabouts of other traffic can be obtained not only visually but also through the use of our auditory channel. Information from both what we hear and what we say can be invaluable in helping to avoid a collision.

Take Advantage of "Party Line" Information

In contrast to today where virtually every telephone is connected to its own dedicated private line—whether it's cellular or land-line—years ago there were party-line telephones where several different subscribers shared the same line. If you accidently (or on purpose!) picked up the receiver when another customer was on the line, you could listen in on their conversations. It was rightfully deemed inappropriate behavior to listen to others' telephone conversations, but in the context of the flight environment the opposite is true—doing so could prevent an accident. Pilots who neglect to listen to **party-line** transmissions made on the same communications frequency by ATC and other pilots are missing valuable information that could help in ascertaining where other aircraft are located. Obviously no benefit is gained if others are not equipped with radios, are not using them as they should, or are not transmitting on the same frequency—all of which are possibilities as discussed in Chapter 1 on runway incursions—but at least you can acquire a better mental picture of the location of those who *are* transmitting on your frequency.

Avoid Finger Trouble

Of course you can't participate in the conversation if you're not on the line. In Chapter 1 we mentioned a malady that affects all pilots at one time or another—finger trouble. While conducting a sightseeing flight along the Hudson River VFR corridor, the pilot of a light airplane—despite making what he thought were all the right position reports on the UNICOM advisory frequency—barely missed (by 200–300 feet) colliding with a helicopter. He later discovered that he had incorrectly entered 123.075 MHz instead of 123.05 MHz in his communication radio (ASRS Report No: 859312).

Sometimes it's the small mistakes that can lead to big trouble, like entering the wrong frequency in the Nav/Com, selecting the wrong switch on the radio/intercom audio panel, or setting an incorrect value in the altimeter subscale. For example, the pilot of a Cessna 172 was issued traffic alerts and instructed to alter heading to avoid a head-on collision; the aircraft's altimeter was incorrectly set to read 1,000 feet below its actual altitude (ASRS Report No: 545368). It's important not to completely trust yourself—or your fellow flight crew member for that matter—when it comes to setting critical values into your radios and other systems. You need to double-check all entries and settings, and if flying in a crewed flight deck, have the other crew member verify their accuracy.

Listen!

> I know you think you understand what you thought I said, but I don't think you realize that what you heard is not what I meant.
>
> —*Author unknown*

Collisions and close calls have occurred because of misunderstood communication. An airplane on final for Runway 26L at Brackett Field in California executed a go-around to avoid a collision with another airplane on final for the same runway. The pilot thought he had heard "26 Left" but was actually cleared to land on Runway "26 Right." During a visual approach to Salt Lake City, two airplanes experienced an NMAC because one of the pilots understood the tower's instruction to descend to "6,500 feet," as "5,500 feet" (ASRS Report No: 867138; 850808).

As we discovered in the previous chapter, communication breakdown is a major factor in accidents and incidents. Just as it is important to double-check the accuracy of frequencies and altimeter settings, it's also important to double-check the content of all communications, whether between you and ATC or you and your fellow crew members. If you have even the slightest doubt, you should seek clarification of what was actually being said. This is a challenge during a "busy" flight or when a radio frequency is congested, but it's better to be safe than sorry.

Make Sure You Are Heard

Effective communication involves a sender who accurately conveys information and a receiver who correctly understands what the sender intended. As the sender, you must not only speak loudly and clearly enough to be heard, but also properly arrange your words and sentences in such a way as to reduce the possibility of misinterpretation by ATC and others. Consult the *AIM* and follow its guidance. It provides advice on radiotelephony technique, such as listening on the frequency to avoid interrupting a conversation already in progress and thinking about what to say before transmitting, using standard and concise phraseology, and using correct phraseology to accurately report your altitude. It also outlines required and recommended procedures to follow, such as accurately identifying yourself on initial call-up and acknowledging all call-ups, clearances and frequency changes. Be especially alert for other aircraft with similar call signs (tail numbers) and make sure air traffic controllers don't inadvertently misidentify you—they have in the past (they are only human, too) and it has led to close calls and even accidents. Accurate communication is even more critical in a non-radar/non-ATC environment; therefore, be sure to follow the procedures outlined in the *AIM* and advisory circulars about announcing your position and intentions at all the recommended locations—especially in the vicinity of nontowered airports.

As the *AIM* reminds us, the single most important goal in pilot-controller communications is understanding, so if ATC or others are having difficulty understanding you—even when you're using standard aviation phraseology—speak in plain English and use whatever words necessary to get your message across.

Follow Rules and Recommended Procedures

Many rules of the air are designed to keep us from flying into each other. For example, there's a reason why aircraft are restricted to flying no faster than 250 knots indicated airspeed (KIAS) below 10,000 feet: Aircraft traveling faster are more difficult for ATC personnel to manage and are harder for pilots to see and steer clear of in time. The rule was adopted after a collision between a TWA Super Connie and a UAL DC-8 over New York City in 1960, which investigators attributed in part to the high speed of the DC-8 (over 300 KIAS). There are several other regulations and recommended procedures that, if followed, will help protect you from experiencing a MAC.

Comply with VFR Weather Rules

Comply with visibility and cloud separation minima; if you are flying VFR, doing so should give you enough room to take evasive action should an IFR aircraft suddenly break out of the clouds. The pilot of a Grumman American Yankee didn't do that; according to the

NTSB report, the aircraft was "skimming" along the base of the cloud deck when a U.S. Air Force Boeing KC-135 Stratotanker broke out of the cloud and immediately collided with the Grumman (NTSB Identification No: LAX82AA106).

Just as your ability to avoid a rear-end collision with another automobile is reduced if you overdrive your headlights at night, so is your ability to maneuver in time to avoid colliding with another aircraft when you overfly your visibility. Therefore, not only should you comply with VFR visibility minimums, but you should also slow down—especially if you fly a fast aircraft—in conditions of reduced visibility.

Adhere to the Cruising Altitudes Rules

Also introduced in the 1960s were the IFR and VFR cruising altitude regulations—sometimes known as the hemispheric, or east-west, rules. During level cruising flight below 18,000 feet above mean sea level (MSL) but above 3,000 feet AGL, when flying on a magnetic course of 360 degrees to 179 degrees, you are required to fly at odd thousand-foot altitudes plus 500 feet when flying VFR (*odd plus 500 eastbound*) and odd thousand-foot altitudes when flying IFR (*odd eastbound*); on a magnetic track of 180 to 359 degrees, you are required to fly at even thousand-foot altitudes plus 500 feet when flying VFR (*even plus 500 westbound*) and even thousand-foot altitudes when flying IFR (*even westbound*). There are exceptions, so consult the appropriate regulation. Also keep in mind that the rule is based on your course (track) over the surface, not heading. For example, if a crosswind from the west causes you to fly a heading of 355 degrees to maintain a course of 004 degrees when flying VFR, you would need to fly at an odd thousand-foot altitude plus 500 feet eastbound (e.g., 5,500 feet MSL, *not* 4,500 or 6,500 feet).

Follow Recommended Traffic Pattern Procedures

One of the earliest fatal collisions involving a civilian airliner occurred in 1947 when a Vultee BT-13 Valiant descended on top of a Delta Air Lines DC-3 at the Muscogee County Airport in Columbus, Georgia, killing everyone on both airplanes. Besides not keeping a diligent lookout, the cause of the accident was the failure of the Valiant pilot to conform to the standard left-hand traffic pattern.[35] Unfortunately, this type of non-compliance with rules and recommended procedures still occurs today. A fatal midair collision between a Mooney M20C and a Cessna 177RG Cardinal took place 0.9 NM from the departure end of Runway 34 at Penticton Regional Airport, in British Columbia. According to the TSB, the Mooney "did not follow recommended circuit-joining procedures" (TSB Report No: A99P0108). More recently, according to the NTSB preliminary report, a Cessna 172 wasn't flying the established left-hand traffic pattern when it collided with a Eurocopter EC35 helicopter at Shenandoah Valley Regional Airport in Weyers Cave, Virginia (NTSB Identification No: ERA11FA101B).

Since most MACs occur in the traffic pattern at nontowered airports, it's crucial that you follow the recommended procedures published in the *AIM* and applicable advisory circulars. Standard traffic patterns and procedures at airports without control towers are designed to create an orderly flow of traffic and reduce the likelihood of a MAC. Pilots who don't adhere to them risk not only their own lives, but the lives of others as well. Before you even get airborne you should know the traffic pattern direction, the proper procedures for entering and exiting the pattern, appropriate altitudes, the correct traffic frequency and call-up points—consult the *A/FD* or the *Canada Flight Supplement* (*CFS*)—and you should comply with them when airborne. Keep in mind that many aircraft still fly without a radio (NORDO), and many accidents have occurred when aircraft have seemingly appeared out of nowhere, so keep a sharp lookout. Finally, if you're planning to fly into any airports north (or south) of the U.S./Canadian border, you should familiarize yourself with the appropriate traffic pattern/circuit entries and departures; despite their similarity, U.S. and Canadian procedures (especially at nontowered/uncontrolled airports) differ significantly from each other—not knowing how could lead to an accident.

Practice the Sterile Cockpit Rule

Several collisions and close calls could have been averted if flight crew members were paying more attention to the job at hand. In the last chapter we pointed out that the aptly named sterile cockpit rule is mandatory for commercial flight operations, and we urged you— whether flying commercially or not—to comply with the rule to reduce the probability of experiencing a collision on the ground. It's also good practice for avoiding a collision in the air. The rule requires you to refrain from nonessential duties or activities during the critical phases of flight (all operations except cruise) below 10,000 feet. Avoiding extraneous conversations and minimizing unnecessary distractions—especially in collision-prone areas such as airports, high-density traffic areas, and navigation aids and waypoints—will increase the effectiveness of your other collision avoidance strategies and your odds of avoiding a collision.

Comply With ATC Clearances

We've already emphasized the wisdom of using ATC as a resource in helping you detect conflicting traffic. The primary purpose of air traffic controllers is to prevent collisions between aircraft, so cooperate with them. In the collision with PSA Flight 182, the Cessna pilot acknowledged and repeated the clearance to maintain a 070 degree heading, but he failed to do so—his actual heading was closer to 090 degrees, the same heading as Flight

182. According to the accident report, had he maintained his assigned heading, he would have cleared the Boeing 727 by about 1,000 feet.

A short refresher on clearances: If ATC issues you a clearance that you don't think is workable, would cause you to deviate from a rule or regulation, or in your opinion would place the aircraft in jeopardy, it is your responsibility to request an amended clearance. However, once you accept a clearance you must adhere to it unless an amended clearance is obtained or an emergency occurs. In addition, if you accept instructions to follow an aircraft or maintain visual separation from it, it is your responsibility to keep your eyes on it and maneuver to avoid it. Should you lose sight of the other aircraft, or are unable to accept the responsibility for separation, you must immediately notify ATC. When the conflict alert warning went off in the San Diego Approach Control facility, the controller wasn't too alarmed since the PSA captain had informed him only moments before that they had the "traffic in sight." Of course we know in hindsight that when they had lost sight of the Cessna, they should not have assumed that it had passed—the captain should have promptly notified the controller.

Practice Caution in Congested and High-Risk Airspace

Several airports in close proximity to one another usually indicate a significant amount of flight activity. For example, there is a high volume of civilian training and military flight operations in the Orlando, Florida, area. Therefore, if there is no need for you to fly through these areas, then don't. It's also recommended that you avoid flying through alert areas (A) (or advisory areas in Canada [CYA]), **military operations areas** (MOA), and **military training routes** (MTR). Flying through these types of **special use airspace** (also known as *Class F airspace* in Canada) increases the probability of experiencing a MAC. In one example, a Cessna 172 and a Piper Seminole—both instructional flights—collided in midair in a concentrated flight training alert area (A-291B) about 18 miles west of Fort Lauderdale–Hollywood International Airport (NTSB Identification No: ERA09FA080).

Outlined by magenta hatched lines on aeronautical charts, alert areas indicate a high volume of pilot training or unusual aerial activity. An explanation of the type of activity will usually be included. In Canada, letters in parentheses after the three-digit designation number are used to indicate the type of activity in an advisory area (e.g., CYA(A), aerobatic; CYA(S), soaring; CYA(T), training). Graphic symbols are also used on charts to indicate different types of aerial activity, including parachute jumping areas, glider operating areas and hang glider activity. If you're not participating in the activity for which the area is designated, it's recommended you avoid it; if you elect to fly through it, you must be especially alert for other aircraft.

MOAs, which consist of airspace of defined dimensions within which military flight activities take place, are also outlined by magenta hatched lines on aeronautical charts. These include air combat tactics and intercepts, aerobatics, formation training, and low-altitude flight. Depicted as gray lines on aeronautical charts, MTRs are also used by the military for the purpose of conducting low-level tactical training. Military aircraft in both areas can operate near the surface up to 10,000 feet MSL (certain MTR segments may authorize higher altitudes) at speeds significantly greater than the 250 KIAS speed limit, leaving little time available for pilots of nonparticipating aircraft to detect and avoid them. Once again, it's recommended to avoid flight into these areas altogether; an encounter with a fast-flying military jet is a quick way to die. If for some reason you are unable to avoid flight into one of these areas, contact any FSS within 100 miles of the area to determine if an MOA or nearby MTR is active (or hot), then contact the controlling agency (frequency listed on the aeronautical chart) for advisories when flying within an active MOA or near an MTR. And remember: *Exercise extreme caution and maintain a vigilant lookout.* Keep in mind that the thin gray line depicting an MTR on the chart is the centerline of the route—the actual width varies for each MTR and may extend for several miles on either side of the charted line. Also, it's wise to never parallel an MTR; you should cross through it as quickly as possible and leave a wide berth.

Finally, as we've mentioned, navigation aids such as VORs attract aircraft like a magnet. There is an even greater—and more deadly—attraction if pilots are flying the centerline using **GPS navigation**. As one pilot discovered when he came within 200 feet of another aircraft passing beside him, the increased accuracy of GPS navigation can reduce the size of an airway to the width of a roadway.[36] If everyone is following the **hemispheric rules** (discussed earlier in this chapter) the chance of encountering another aircraft at your altitude is minimal; however, the rules don't protect you from an overtaking (or an overtaken) aircraft traveling in the same general direction. Also, there's always a chance that opposite direction traffic may be at your altitude; that's what happened in the Gol Airlines B-737 and Embraer Legacy collision in Brazil in 2006. Through a series of errors, both airplanes, flying in opposite directions on the same airway and under the control of ATC, ended up cruising at the same altitude at flight level 370 (FL370 or 37,000 feet). Since horizontal navigation was strictly maintained by both aircraft, many have observed that the precise accuracy of the modern onboard GPS-based navigation systems increased the likelihood of the collision.

If you fly in international oceanic or remote airspace—where there is a lack of ATC ground radar to provide positive separation—you should follow the guidance provided by ICAO and the FAA to reduce the risk of a collision that could arise from highly accurate navigational systems. Pilots are encouraged to use published **strategic lateral offset procedures** (SLOP) to reduce the threat of a collision with an aircraft that has deviated from its altitude, and to mitigate encounters with wake turbulence. Assuming your navigation equipment has automatic offset programming capability, you have the option of flying laterally offset the route centerline by one NM or two NM to the right. See the FAA Advisory Circular 91-70, *Oceanic and International Operations*, for more information. Unfortunately, for VFR pilots there is no such guidance on how to minimize your odds of a collision while navigating on airways in domestic airspace, except to keep a sharp lookout and avoid flying directly overhead VORs or in their vicinity.

Use Correct Procedures for Onboard Collision Avoidance Technology

The transponder in the ERJ Legacy was off when it collided with the Gol Airlines B-737. The NTSB—a party to the Brazilian investigation—concluded that this was likely due to inadvertent deactivation by the flight crew. This not only precluded ATC from discovering the aircraft was at the wrong altitude, but since TCAS is rendered inoperative if the transponder is off or in standby mode, it also failed to warn the crew of the impending collision. Therefore, if you fly TCAS-equipped aircraft, it is critical that you ensure the transponder is properly activated during your flight. You should also be thoroughly familiar with how to interpret traffic display symbology, including shapes, colors and numerics; the appropriate response to visual and aural voice-command alerts (TAs and RAs); and proper communication procedures with ATC—especially if a TCAS alert requires a deviation from an ATC clearance.

If you fly for an air carrier, you will receive TCAS instruction during initial, transition and upgrade training. Information will also be published in the aircraft manufacturer's AFM or operating manual or your company SOPs. The FAA also publishes guidance in the *AIM* and in Advisory Circular 120-55 outlining ways in which flight crews can effectively operate TCAS and respond to alerts.

Finally, don't rely on TCAS as your sole traffic collision avoidance tool—despite its success, it is not a perfect system. For example, TCAS cannot detect aircraft that are not equipped with functioning transponders. Also, the older TCAS I systems provide proximity warnings only; they do not provide recommended avoidance procedures.

Most GA aircraft are not equipped with TCAS, but some are beginning to be furnished with the promising new technology of automatic dependent surveillance-broadcast (ADS-B). Originally initiated in Alaska in 2001 as part of the Capstone Program, ADS-B coverage is gradually expanding to the rest of the United States, and ADS-B avionics will be required by the year 2020 for aircraft operating in most controlled airspace requiring transponders.[37] Billed as the backbone of the FAA's NextGen (next generation) system that will modernize the NAS by gradually moving away from ground-based radar surveillance and navigation systems to more GPS satellite-based technology, ADS-B uses inputs from both aircraft and ground stations to provide ATC and pilots with a more accurate picture of where aircraft are located. Pertinent to our discussion of collision avoidance is the traffic information service-broadcast (TIS-B) component of the system. ADS-B equipped aircraft broadcast their identification, position, altitude, speed and track every second to other aircraft and to compatible ground-based transceivers (GBTs)—known as *ADS-B Out*. In turn, these aircraft receive similar information directly from other appropriately equipped aircraft and indirectly from GBTs that retransmit ATC radar traffic information—known as *ADS-B In*. If you are flying an ADS-B equipped aircraft you will see the location, altitude and direction of other nearby aircraft on your onboard cockpit display of traffic information (CDTI).[38]

It's important to understand that as of this writing, ADS-B/TIS-B is *not* to be used as a collision avoidance system—the technology and infrastructure is still in development and not all aircraft can be displayed on the CDTI. It's only intended to assist you in visually acquiring other aircraft; you are still responsible to *see and avoid*, and if you tell ATC you have "traffic in sight," it must be based on you actually *seeing* the aircraft out the window and not on information depicted on the CDTI. Consult the current *AIM* for more up-to-date information.

Avoid Flying in "Formation"

Finally, a word to those who have never engaged in formation flying, but would like to: Don't. Eight of seventeen in-flight collisions in Canada during a 10-year period involved some type of formation flying—that's almost half![39] If the risk of dying in a midair is considerably elevated for well-trained and highly-skilled aerobatic and military pilots engaged in this activity, why would you want to risk it? Flying in formation is no sport for amateurs—leave it to the professionals.

We've made great strides in reducing the number of MACs in the years since the Grand Canyon accident: extensive radar coverage and enhanced ATC facilities, tighter rules and regulations, sophisticated technology in the cockpit, and better procedures and training have all contributed to this success. However, even though they happen relatively infrequently compared to other types of accidents, in-flight collisions still occur and are still deadly. In many ways we've come a long way since that tragic accident that claimed the lives of 128 people over 50 years ago, but in some respects we haven't: When flying VFR, or IFR in VMC, or when asked by ATC to maintain visual separation, you as the pilot in command are *still* responsible for avoiding collisions with other aircraft.

As a pilot you know there are a few basic, yet fundamental, top-priority goals that you must achieve if you're going to have a successful flight: one is to avoid stalling, and the other is to avoid hitting anything. Follow the strategies suggested in this chapter to accomplish the latter: When in VMC don't get buried in the cockpit, but instead spend most of your time looking outside and scanning regularly for other traffic. Solicit others' help in your visual search and help other pilots and ATC see you better by turning on your landing lights—especially in high-density traffic areas—and correctly activating your transponder. Carefully listen in on others' transmissions to form a more accurate mental picture of their locations and use effective communication strategies to let them know where you are. Take advantage of ATC traffic advisory services to help you spot others and inform ATC when you're unable to see the traffic. Comply with ATC instructions and let the controller know when you're unable to. Understand and use onboard collision avoidance technology appropriately, follow regulations designed to separate you from other traffic, and practice recommended procedures designed to reduce your chance of a collision—especially at nontowered airports. Most importantly, always be vigilant. Never let your guard down by becoming complacent—especially in areas prone to in-flight collisions. Remember the old adage: *Just because you're not paranoid, doesn't mean they're not out to get you.*

Helpful Resources

Pilots' Role in Collision Avoidance (FAA AC 90-48). Though due for updating, this advisory circular provides information on common causes of midair collisions and gives guidance on ways in which you can reduce your chance of experiencing one. Figure 2-2 is also found in this document. (www.FAA.gov/regulations_policies/advisory_circulars/index.cfm/go/document.information/documentID/23090)

Vision in Aviation: To See or Not to See. A 16-minute video produced by the Airman Education Programs branch at the FAA's Civil Aerospace Medical Institute (CAMI) that focuses on the visual limitations of flight, including those involved in MACs. (www.FAA.gov/about/office_org/headquarters_offices/avs/offices/aam/cami/library/online_libraries/aerospace_medicine/aircrew/physiologyvideos/english/)

The AOPA Air Safety Institute has two free online interactive courses that can increase your understanding of various types of airspace to help you avoid the threat of collision. (www.AOPA.org/asf/online_courses)
- *Know Before You Go: Navigating Today's Airspace*
- *Mission Possible: Navigating Today's Special-Use Airspace*

AOPA Air Safety Institute's *Safety Advisors* can assist you in understanding and reducing the threat of a midair collision. (www.AOPA.org/asf/publications/advisors.html)
- *Collision Avoidance*
- *Operations at Nontowered Airports*

Characteristics of U.S. Midairs, published in the May/June 2001 edition of *FAA Aviation News*, carefully examines the characteristics of 329 midair collisions that took place in the United States between 1983 and the summer of 2000. (www.FAA.gov/news/safety_briefing/2001)

The SeeAndAvoid.org portal provides civilian and military pilots with information and education on airspace, visual identification, aircraft performance and mutual hazards to safe flight, with the ultimate goal of eliminating midair collisions and close calls through good flight planning. (www.SeeAndAvoid.org)

Notes

1. Civil Aeronautics Board, *Accident Investigation Report: Trans World Airlines Inc., Lockheed 1049A, N69020, and United Air Lines Inc., Douglas DC-7, N63240, Grand Canyon, Arizona, June 30, 1956,* File No. 1-0090 (Washington, DC: CAB, April 15, 1957).

2. International Civil Aviation Organization, *News Release: Growth in Civil Aviation, 1968: World Air Traffic Achieves Another Good Year in the Face of Difficult Circumstances* (Montreal, Canada: ICAO, December 23, 1968).

3. National Transportation Safety Board, *Briefs of Accidents Involving Midair Collisions: General Aviation 1978,* NTSB-AMM-80-2 (Washington, DC: NTSB, May 20, 1980).

4. National Transportation Safety Board, *Midair Collisions in U.S. Civil Aviation—1968: A Special Accident Prevention Study,* NTSB-AAS-69-AA, Appendix 9: "Chronology of the Airline Search for a Collision Avoidance System" (Washington, DC: NTSB, July 1969): 79.

5. Federal Aviation Administration, *Airworthiness Approval of Traffic Alert and Collision Avoidance Systems (TCAS II) Version 7.0 and Associated Mode S Transponders,* AC 20-151 (Washington, DC: FAA, February 7, 2005).

6. Federal Aviation Administration, *Introduction to TCAS II, Version 7.1* (Washington, DC: FAA, 2011).

7. Transport Canada, *Regulations for Airborne Collision Avoidance System* (Ottawa, ON: TC, 2007).

8. Robert C. Matthews, "Characteristics of U.S. Midairs," *FAA Aviation News* 40.4 (May/June 2001): 1–3.

9. U.S. Bureau of Transportation Statistics, *No. of Pilot-Reported Near Midair Collisions (NMAC) by Degree of Hazard* (Table 2-15) (Washington, DC: U.S. Bureau of Transportation Statistics). Available online at www.RITA.dot.gov/bts/sites/rita.dot.gov.bts/files/publications/national_transportation_statistics/index.html#chapter_2.

10. Boeing Commercial Airplanes, *Current Market Outlook 2013–2032* (Seattle, WA: Boeing, Summer 2013). Available online at www.Boeing.com/commercial/cmo.

11. Federal Aviation Administration, *FAA Aerospace Forecasts: Fiscal Years 2013–2033* (Washington, DC: FAA, 2012).

12. National Transportation Safety Board, *Special Study: Midair Collisions in U.S. Civil Aviation, 1969–1970,* NTSB-AAS-72-6 (Washington, DC: NTSB, June 7, 1972).

13. AOPA Air Safety Foundation, *2001 Nall Report: General Aviation Accident Trends and Factors for 2000* (Frederick, MD: AOPA, 2002).

14. AOPA Air Safety Foundation, *Safety Advisor: Operations at Nontowered Airports,* Operations & Proficiency No. 3 (Frederick, MD: AOPA, January 2007).

15. "Mid-air Collision," *Aviation Safety Letter,* Issue 2/2001 (Spring 2001): 1–3.

16. Australian Transport Safety Bureau, *Review of Midair Collisions Involving General Aviation Aircraft in Australia between 1961 and 2003,* Research Report B2004/0114 (Canberra City, Australia: ATSB, May 2004).

17. AOPA Air Safety Foundation, *Safety Advisor: Collision Avoidance Strategies and Tactics,* Operations & Proficiency No. 4 (Frederick, MD: AOPA, August 2006).

18. Narinder Taneja and Douglas A. Wiegmann, "Analysis of Mid-Air Collisions in Civil Aviation," *Proceedings of the 45th Annual Meeting of the Human Factors and Ergonomics Society* (Santa Monica, CA: Human Factors & Ergonomics Society, 2001).

19. National Transportation Safety Board, *Annual Review of Aircraft Accident Data U.S. General Aviation, Calendar Year 2005*, NTSB/ARG-09/01 (Washington, DC: NTSB, May 26, 2009).

20. AOPA Air Safety Foundation, *Flight Instruction Safety: An In-Depth Look at Instructional Accidents* (Frederick, MD: AOPA, July 2004).

21. Bruce Landsberg, "Formation Fumbles," *AOPA Online* 44.7 (July 2001). Available online at www.AOPA.org/asf/asfarticles/2001/sp0107.html.

22. Matthews, "Characteristics of U.S. Midairs," 1–3.

23. J.W. Andrews, *Unalerted Air-to-Air Visual Acquisition*, Report No. DOT/FAA/PM-87/34 (Lexington, MA: MIT Lincoln Laboratory, November 26, 1991).

24. Ibid.

25. "Unalerted 'See & Avoid' Not Good Enough," *Aviation Safety Letter* Issue 1/95 (Winter 1995): 1-2.

26. H.W. Leibowitz and D.A. Owens, "Anomalous Myopias and the Intermediate Dark Focus of Accommodation," *Science* 189 (1975): 646–648.

27. Federal Aviation Administration, *Pilots' Role in Collision Avoidance*, AC 90-48C (Washington, DC: FAA, March 18, 1983).

28. AOPA, *Safety Advisor: Collision Avoidance*

29. Ibid.

30. Kurt Colvin, Rahul Dodhia, and R. Key Dismukes, "Is Pilots' Visual Scanning Adequate to Avoid Mid-air Collisions?" *Proceedings of the 13th International Symposium on Aviation Psychology* (2005): 104–109.

31. Ibid.

32. Matthews, "Characteristics of U.S. Midairs," 1–3.

33. Federal Aviation Administration, *Aviation Instructor's Handbook*, FAA-H-8083-9A (Washington, DC: FAA, 2008).

34. J.W. Andrews, "Modeling of Air-to-Air Visual Acquisition," *The Lincoln Laboratory Journal* 2.3 (1989): 475–480.

35. Civil Aeronautics Board, *Accident Investigation Report: Delta Air Lines Inc., Columbus, Georgia, April 22, 1947* (Washington, DC: CAB, August 1, 1947).

36. "GPS En Route Navigation" (Letter to Editor), *Aviation Safety Letter* Issue 2/96 (Spring 1996): 5–6.

37. Federal Aviation Administration, "14 CFR Part 91, Automatic Dependent Surveillance-Broadcast (ADS-B) Out Performance Requirements to Support Air Traffic Control (ATC) Service; Final Rule," *Federal Register* 75.103 (May 28, 2010): 30160–30195.

38. Federal Aviation Administration, *Airworthiness Approval for ADS-B In Systems and Applications*, AC 20-172 (Washington, DC: FAA, February 28, 2011).

39. "Mid-air Collision," *Aviation Safety Letter,* 1–3.

Do You Want to Be a Test Pilot?

3

They didn't expect it. One minute the Embraer EMB-120 Brasilia, flying on autopilot in instrument meteorological conditions (IMC), was level at 4,000 feet in a left turn on radar vectors to intercept the localizer for an instrument approach on Runway 03R at Detroit Metropolitan/Wayne County Airport; the next minute it was in a 140-degree left bank and a 50-degree nose-down attitude. The flight crew fought to regain control of the airplane, but there wasn't enough altitude. Comair Flight 3272, carrying three crew members and 26 passengers, struck the ground in a steep nose-down attitude killing all aboard. What was the cause of this dreadful accident? Ice. According to the NTSB, the airplane lost control and stalled because of an accumulation of a thin rough accretion of ice on its lifting surfaces (Report No: NTSB/AAR-98/4).

Two years earlier, the crew members of another twin-engine turboprop transport—American Eagle Flight 4184—lost control of their ATR-72 due to uncommanded roll excursions while descending to 8,000 feet in a holding pattern. They too tried unsuccessfully to regain control. Less than a minute after the first roll upset, all 68 passengers and crew members were killed as the airplane crashed in a field near Roselawn, Indiana. The culprit in this horrific accident also was **airframe icing** (Report No: NTSB/AAR-96/01).

A quick glance at the statistics indicates the deadly nature of this hazard. NTSB researchers conducted a study of icing accidents that occurred over a 19-year period ending in 2000; they found at least 583 airframe icing accidents in the United States resulting in more than 800 deaths.[1] Over the 10 years (1998–2007) since the Comair accident, the NTSB investigated 264 airplane icing accidents that were responsible for the loss of 202 lives.[2] You can add about two dozen more in the few years since then, including the February 2009 crash of a Colgan Air/Continental Connection Dash 8, near Buffalo, New York, that claimed the lives of 50 people (Report No: NTSB/AAR-10/01).

Most airframe icing accidents in the United States involve GA aircraft, with most of these accident aircraft not certified for flight into icing conditions. In fact, four out of five airframe icing accidents reported in the NTSB study cited earlier involved GA aircraft, and they were responsible for the loss of 522 lives. The remaining 20 percent involved Part 135 and 121 commercial flight operations, but they were responsible for more than a third of all icing-related deaths (297, or 36 percent of fatalities). This high percentage, of course, is due to the fact that when a large passenger-carrying airplane goes down, it takes more people with it than does a smaller one. However, even if you are an experienced pilot, be aware that although many airframe icing accidents involve pilots with lower flight experience—40 percent had fewer than 1,000 flight hours—one in four pilots had more than 3,000 hours of flight experience, almost one-fifth held an airline transport pilot (ATP) certificate, and 80 percent possessed an instrument rating.[3] In addition to the three tragic accidents mentioned above, Table 3-1 lists some of the most notable fatal airframe icing accidents in the United States and Canada involving experienced and professional flight crews flying transport category aircraft; a considerably longer list could be made of similar accidents worldwide.

Date	Operator	Aircraft	Location	Fatalities
Feb 16, 2005	Circuit City Stores	Cessna Citation	Pueblo, CO	8
Nov 28, 2004	Global Aviation	Canadair CL-600	Montrose, CO	3
Apr 8, 2003	Grand Aire Express	Falcon 20	Toledo, OH	3
Mar 22, 1992	USAir	Fokker F-28	LaGuardia, NY	27
Feb 17, 1991	Ryan International	DC-9	Cleveland, OH	2
Dec 26, 1989	United Express	BA Jetstream 31	Pasco, WA	6
Mar 15, 1989	Mid Pacific Airlines	NAMC YS-11	W. Lafayette, IN	2
Mar 10, 1989	Air Ontario	Fokker F-28	Dryden, ON	24
Nov 15, 1987	Continental Air Lines	DC-9	Denver, CO	28
Dec 12, 1985	Arrow Air	DC-8	Gander, NF	256
Jan 13, 1982	Air Florida	B-737	Washington, DC	78

Table 3-1. Partial list of fatal U.S. and Canadian airframe icing accidents.

Clearly, aircraft icing is no respecter of flight time; it strikes both novice and veteran pilots alike. Four months after the 1997 Comair Flight 3272 accident, the NTSB put aircraft structural icing on its Most Wanted List of needed safety improvements.[4] Although much has been done over the years to reduce this hazard, as late as 2011, it was still on the list. Why? Because pilots are still being caught by it.

Aircraft Icing

There are generally two types of aircraft icing: induction icing and airframe (or structural) icing. **Induction icing** reduces the amount of air entering an engine's induction system, reducing its power output. A common type of reciprocating-engine induction ice is carburetor ice, where moisture freezes inside the intake of a float-type carburetor from cooling caused by vaporization of fuel and the expansion of air. Induction icing is one of the most common causes of engine failures in reciprocating-engine-powered GA aircraft and is a hazard that pilots must manage. However, this chapter focuses on how to manage the threat of ice that can accumulate on the wings, tail, control surfaces and other parts of the airframe.

Meteorology of Icing

To understand how an iced-up aircraft affects your ability to fly safely, you must first know how it forms. The two essential requirements for its creation are freezing temperatures and moisture. The outside air temperature (OAT) doesn't necessarily have to be at or below freezing—it can be 5°C or even warmer—but the skin of the aircraft does. In addition, the formation of airframe ice almost always requires the presence of visible moisture (e.g., clouds, fog or precipitation), but occasionally invisible moisture in the form of water vapor and high relative humidity will be enough. With the exception of large hail (abbreviated in aviation weather reports and forecasts as GR), which can cause substantial damage to an aircraft's skin, water remaining in solid form (e.g., ice crystals, IC; snow, SN; snow grains, SG; ice pellets, PL) usually poses no harm when it strikes an aircraft; it usually flows around or bounces off the airframe. It is water in liquid (or partially liquid) form that actually causes trouble. Oddly enough, water can exist in the liquid state even when the temperature is below freezing. These water droplets are said to be supercooled. In fact, in controlled laboratory conditions, they can remain in the liquid state to temperatures as low as -40°C. However, in supercooled clouds—where most in-flight icing occurs—the proportion of liquid water droplets to ice crystals reduces as the temperature drops, until at temperatures below about -20°C clouds are mostly made up of ice crystals. For supercooled water droplets to freeze, they need something to freeze onto such as dust or salt particles, ice crystals, or the frigid surface of an aircraft. Once they strike the aircraft, they freeze, forming ice that clings to the wings, control surfaces and other critical parts of the airframe.

Rate of Catch

The severity of ice accumulation is proportional to the quantity and size of the supercooled water droplets and the speed at which an aircraft flies through them. An airfoil's shape also affects the **rate of catch**; airfoils with thin knife-edged leading edges—such as those on smaller aircraft and on the wingtips of tapered or swept wings—catch more supercooled water drops than thicker ones do. Likewise, thin tail surfaces catch more supercooled water drops, sometimes resulting in more ice on the hard-to-see tail than on the main wing. It may seem odd that a thinner leading edge would attract more ice, but the reason has to do with the airflow around the wing. There is an area of high positive pressure near the leading edge of the airfoil called the stagnation point where airflow velocity is zero. Since the stagnation point for thin airfoils is smaller and the positive pressure is lower, it is less able to deflect droplets around it: They fly right through the stagnation point and adhere to the surface. Stagnation pressures associated with large thick airfoils are high enough to counter the inertia of supercooled water droplets and are therefore better at deflecting them around and past the wing.

Effects of Airframe Icing

Airframe icing leads to a variety of troubling consequences, including reduced visibility through the windscreen, blocked pitot/static vents that cause erroneous flight instrument readings, blocked engine and carburetor air intakes, frozen control surfaces, propeller vibration, vibration or breaking of radio antennas, and compressor stalls in turbine engines from ingested ice that has shed off the airframe. That sounds bad enough, but the major effect of airframe icing is significantly degraded flight performance. Ice accretion disrupts the smooth flow of air over lifting surfaces and changes their aerodynamic shape, resulting in an increase in drag and weight, and a decrease in lift and thrust. According to the Federal Aviation Administration (FAA) Advisory Circular (AC) 91-74, *Pilot Guide: Flight in Icing Conditions*, exposure to certain types of ice, even for short durations, can reduce the coefficient of lift of an airfoil by as much as 50 percent and increase drag by as much as 200 percent![5] Not only does this affect takeoff performance—many pilots have attempted takeoffs with ice contamination and never made it much farther than the end of the runway—but climb performance, cruise speed, service ceiling and maximum range are all reduced.

Airflow disruption over the wings also decreases the ability to maintain control of the aircraft, and early airflow separation decreases an airplane's stalling angle of attack (AOA) and increases its stalling airspeed. Pilots who were unable to quickly extricate themselves from a serious icing encounter have reported that to maintain a safe airspeed, they had no choice but to keep the airplane in a descent with full power; if they tried to remain level, their aircraft would stall. This is what the pilot of a Cessna 208 Caravan discovered. After encountering ice, the Part 135 cargo flight couldn't maintain an altitude of 7,500 feet or airspeed any faster than 125 knots (about 5 knots above the stall buffet speed). The pilot was forced to descend and divert to the nearest available airport. With full power, and at an airspeed just above the stall, the pilot barely made it to a closed runway after stalling 10 feet above the threshold during the landing flare (ASRS Report No: 881246).

Types of Airframe Icing

There are generally three major types of airframe icing: rime icing (RIME ICG), clear icing (CLR ICG), and frost (FRST).

Rime Ice

Rime ice consists of small supercooled water droplets that usually form in clouds or in freezing fog (FZFG). Commonly found well above the freezing level in stratiform-type clouds that form in stable air, these small supercooled water droplets tend to freeze instantaneously, creating an opaque, whitish, brittle ice mostly on the leading edges of aircraft surfaces. Rime ice is usually easier to remove with deicing equipment than clear ice.

Clear Ice

Clear ice—sometimes called glaze ice—consists of large supercooled water droplets that form in clouds or freezing precipitation (freezing drizzle, FZDZ; freezing rain, FZRA). In clouds, these droplets are mostly found just above the freezing level. They are present in layered stratiform-type clouds, but since they are large, they are also more prevalent in conditions that support updrafts: cumulo-type clouds in unstable air and on the windward side and crests of hills and mountains. Their large size also means they release more latent heat when they freeze than rime ice-sized droplets do. Latent heat is the heat energy released into the atmosphere when a liquid droplet turns into ice. It is equal to the amount of energy required to keep the water in liquid state and is stored in the form of kinetic energy. Since the temperature is close to freezing, the release of latent heat as the droplet freezes momentarily raises the temperature above freezing, causing it to melt and flow backward on the airplane's surface before refreezing (called run-back icing). Usually more hazardous than rime, clear ice is very hard, glassy and relatively transparent. It flows farther aft of the leading edges and is more difficult to remove with deicing equipment than rime ice.

It's important to note that clear ice and rime ice are sometimes found together. As for rime and clear ice, the severity of **mixed ice** (MX ICG) depends on its quantity, location, shape and roughness. Mixed ice can often form unusual shapes, leading to significant airflow disturbance over the airframe.

Frost

Frost usually forms on an aircraft that is parked outside on clear winter nights when the air cools to the frost point (dew point temperature below 0°C). In this case, water vapor sublimates directly from gas into ice without first changing to liquid. Frost primarily disrupts the smooth flow of air over the wing and other lifting surfaces. It can also form on a cold-soaked aircraft where the wings have been cooled to below-freezing temperatures—even when the OAT is well above 0°C—from proximity to cold fuel stored in the wings' fuel tanks. This can occur if the tanks contain a large quantity of fuel (e.g., full tanks or extra fuel being tankered) and the aircraft has been flying for a sufficient length of time at subfreezing altitudes. If enough water vapor (high relative humidity) is present after landing, frost—called **cold-soaked fuel frost** (CSFF)—can form. If liquid water or wet snow is present, it can freeze on the aircraft's surfaces. Cold-soaking has contributed to a number of takeoff icing accidents, including an Air Ontario Fokker F-28 accident in Dryden, Ontario, in which 24 people died after the crew attempted to take off with wet snow on the wings. The cold-soaked fuel in the wing tanks cooled the surface of the wings, and the snow subsequently froze.[6]

Though relatively rare, if the airframe is cold enough, frost can also form in flight after descending from a subfreezing altitude to a very humid above-freezing altitude. In this case, the frost is usually short-lived.

Recent Icing Research

In the wake of the American Eagle Flight 4184 accident, a number of researchers and organizations—including NASA's Glenn Research Center (GRC), Environment Canada and the FAA—began to take a closer look at the problem of airframe icing. As a result of laboratory and in-flight icing tests, meteorological and climatological studies, as well as a closer examination of airframe icing accidents, we now have a better understanding of the nature of this hazard.

Supercooled Large Droplets

One important result of this research is a greater appreciation of the hazard of **supercooled large water droplet** (SLD) icing. According to FAA Advisory Circular 91-74, most aircraft icing occurs in clouds that contain supercooled cloud droplets that have an average diameter of about 20 micrometers (μm). However, accident investigators discovered that along with the presence of SLD, American Eagle Flight 4184 also encountered supercooled drizzle and rain droplets the size of which exceeded the criteria in the FAA icing certification envelope (Report No: NTSB/AAR-96/01). The aircraft was flying in an icing environment that contained SLD that ranged between 100 μm and 2,000 μm in diameter. In contrast, the U.S. aircraft icing certification standards—found in 14 CFR §25.1419, Appendix C—allow for a maximum mean effective diameter of only 40 μm.

Supercooled large droplets have a diameter greater than 50 μm and consist of either FZDZ (diameter ≥50 μm to ≤500 μm) or the largest of drops—FZRA (diameter >500 μm). Figure 3-1 illustrates the difference in size between a typical cloud droplet and a typical FZRA droplet—the mass of a typical FZRA droplet is 1,000,000 times that of a typical cloud droplet!

A common weather condition that fosters the creation of SLD icing is a temperature inversion, where FZDZ or FZRA forms after rain (RA) or drizzle (DZ) falls from a layer of above-freezing warm air aloft into a cold subfreezing layer below. **Temperature inversions** are commonly associated with winter warm fronts or stationary fronts where warm air overruns cold air, creating SLD icing conditions that typically cover a large geographical area. Though limited horizontally, icing associated with cold fronts often extends to higher altitudes and can build up rapidly due to the vertical updrafts that are conducive to sustaining the weight of large supercooled droplets. Freezing precipitation in the form of FZDZ and FZRA—the most severe form of clear ice and sometimes called **icing in precipitation** (ICGIP) by aviation weather personnel—poses a serious threat to any aircraft flying underneath a cloud deck. Since supercooled large droplets are substantially larger, and the release

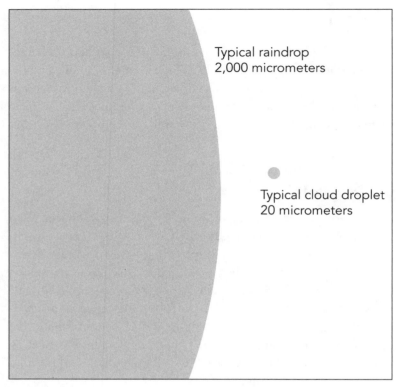

Figure 3-1. Relative size of typical cloud and rain drops.[7]

of latent heat is significantly greater than in the average cloud-sized droplets, for aircraft equipped with **ice protection systems** (i.e., deice and anti-ice equipment), run-back icing can extend beyond the ice-protected regions of the airframe (e.g., farther aft of pneumatic deice boots) creating a considerable hazard.

However, as discovered in the Roselawn accident and confirmed in the latest icing research, SLD is not only confined to precipitation below clouds; it can also exist as large droplets within clouds independent of any temperature inversion. More common than once thought, SLD—primarily FZDZ—is created when smaller supercooled droplets collide with each other and coalesce to form larger droplets—a phenomenon known as the collision-coalescence process. So even if your aircraft is certified for flight into icing conditions, SLD icing can pose a significant threat even when flying in clouds.

Thin Rough Ice

Recent research indicates that it's not always the quantity of ice that determines the degree of airplane performance degradation; the shape, roughness and location of ice are sometimes even more critical. Conventional wisdom suggests that the larger the size of SLD and the greater its quantity, the worse the airframe icing, but this is not always the case. Studies have found that a very thin, yet rough layer of ice can sometimes pose an even greater risk than SLD icing. Since SLDs release significant amounts of latent heat, the run-back icing produced when they freeze often forms a relatively smooth coating of ice that more or less conforms to the shape of the aircraft's surface—it doesn't form the sharp rough edges that can significantly disrupt the smooth airflow over the wings. However, small amounts of ice—even ice less than one-eighth inch thick created by smaller supercooled droplets—sometimes creates irregular or jagged shapes, or a very coarse sandpaper-like texture, that can drastically disrupt the smooth airflow over the lifting surfaces.[8] Therefore, pilots who've had previous success in carrying large amounts of clear smooth ice on their wings may be lulled into a false sense of security when they encounter small amounts of thinner rough ice.

Ice Ridging

Airframe icing accidents usually result from a wing stall. However, experts have discovered that these accidents can occasionally be caused by an uncommanded or uncontrolled **roll upset**. This was the case for American Eagle Flight 4184. According to the NTSB, the loss of control was not the result of a wing stall, but a sudden and unexpected aileron hinge moment reversal that occurred because of airflow separation in front of the ailerons. This in turn was caused by a **ridge of ice** that had built up aft of the trailing edge of the deice boots.

Wind tunnel tests conducted by NASA's GRC Icing Branch have replicated this phenomenon. Because of the large droplet size and greater inertia, SLD ice can flow farther aft of the ice-protected regions of the wing. Since it cannot be removed by leading-edge thermal heating or pneumatic deice systems, it builds up and forms a ridge of ice that disturbs the airflow in front of the ailerons. If the ridge grows large enough, it can cause complete airflow separation in front of the aileron, leading to total loss of roll control. Aircraft that are equipped with pneumatic deice boots and unpowered roll-control systems are particularly susceptible to this phenomenon. Since unpowered roll-control systems are not equipped with hydraulic or electrical boost power to assist in the movement of the flight controls, pilots are less able to exert the physical force necessary to overcome any uncommanded rolling moments.[9]

Ice Bridging

Until recently, many aircraft manufacturers specified that pilots wait until airframe ice accumulates to a minimum thickness (typically one-fourth to one-half inch) before activating the pneumatic deice boots. The reason for this was to avoid the phenomenon known as **ice bridging**. It was believed that if the boots were inflated too soon, when the ice was too thin and pliable, the ice would expand with the inflating boot without breaking and remain over the boot after it deflated, leaving a void of air—in the shape of an inflated deice boot—between the boot surface and the ice. This would render subsequent boot inflations ineffective as ice continued to accrete on this bridge of ice. Waiting for the ice to build up to a minimum thickness before activating the boots ensured that the ice was hard enough to crack and break off, precluding the possibility of ice bridging.

However, recent research has found that incidents of ice bridging reportedly occurred on older deice boot systems (mostly on piston-powered aircraft) equipped with larger tubes that inflate at lower pressures, resulting in lower inflation and deflation rates. A small number of known cases are confined to designs that date back more than a quarter century ago. But the research indicates ice bridging doesn't usually occur on newer turbine-powered aircraft that use smaller tubes, higher inflation pressures, and faster inflation and deflation times. Furthermore, a search of the NTSB's accident database found no accidents in the United States related to ice bridging.[10] In contrast, testing indicates that activating the pneumatic deice boots as soon as ice is detected will effectively prevent ice from accumulating on wing leading edges and will not create ice bridging. There will likely be residual ice—called **inter-cycle ice**—that may not shed during the first inflation cycle, but it will eventually shed during subsequent cycles. Residual ice is more common as both airspeed and OAT decrease.

Tailplane Ice

Though relatively rare, accidents and incidents have occurred due to **ice-contaminated tailplane stalls** (ICTS). This phenomenon was responsible for at least 16 accidents in the United States over a 30-year period, claiming the lives of 139 people, and it is suspected to have played a role in several other accidents.[11]

The center of lift (or pressure) on a wing for most airplanes is aft of the center of gravity (CG), causing a nose-down moment; the horizontal stabilizer—a mini wing, only upside down—provides the tail-down force (or negative lift) necessary to maintain balance along the longitudinal axis (pitching). Airplanes are usually designed so the stalling AOA of the tail is never reached during flight; however, if the tail is sufficiently contaminated with ice, a reduction of its downward lift could result from the ice-induced reduction in its stall AOA. This could lead to erratic elevator movement or, at worse, a tailplane stall, causing a sudden and uncontrollable nose-down pitching movement. Since the downwash angle of the main wing increases and the negative AOA of the horizontal tail surface increases with trailing edge flaps, tailplane stalls are most likely to occur while extending the flaps to the landing position or with the flaps already extended to that position while flying in icing conditions. Unfortunately, the phase of flight most often associated with flap extension is one that is closest to the terrain—the approach and landing.

Airframe Ice Detection

Several accidents have occurred because pilots were unable to accurately detect the presence of ice on their airframe. For example, the NTSB determined that the crew members of a USAir F-28, which crashed after departing Runway 13 at New York's LaGuardia Airport, could not from their respective seating positions adequately ascertain the degree of ice accumulation on the upper surface of the wings (Report No: NTSB/AAR-93/02). Icing experts agree that on some aircraft it is difficult, if not impossible, to visually determine the presence of ice from the cockpit—especially clear ice during the hours of darkness. After experiencing control difficulties on final approach, the crew members of a large turbojet aircraft conducted a physical inspection of the aircraft after landing; they were astonished to find that clear ice almost covered their entire airplane! However, the only ice they could see from their cockpit seats was that which had formed on the wiper blades and attaching bolts (ASRS Report No: 466880).

Ultimately, the answer lies in equipping aircraft with a reliable system that detects and either automatically removes ice on the critical surfaces of the aircraft or alerts the flight crew when to do so. Fortunately, this is part of recent FAA-mandated changes to the airworthiness standards for type certification of new transport category airplanes certified for flight in icing conditions. The regulation requires either the installation of a **primary ice detection system** or changes to the AFM to ensure timely activation of the airframe **ice protection system** (IPS).[12]

Managing the Icing Hazard

Airframe icing is an ever-present hazard in aviation. It poses a threat not only to those who fly aircraft not certified for flight into icing conditions, but also to those who fly aircraft equipped with IPS, as accidents and research have confirmed. Whatever type of aircraft you fly, the rest of this chapter provides information on how you can identify, avoid and escape from an icing encounter. If you fly an aircraft equipped with ice protection, it also provides suggestions on how you can safely manage the icing threat if there is no quick way of escape.

Know Before You Go

As with most hazards, managing airframe icing begins before leaving the ground. Airman's meteorological information (AIRMETs) and significant meteorological information (SIGMETs) are the primary sources for determining icing conditions for your flight. Consult them during your preflight weather briefing to determine freezing levels and areas of moderate to severe icing and freezing precipitation. You can also supplement these with information from the current icing and forecast icing products (CIP/FIP) or the turbulence, icing and freezing level forecast that is part of the *Graphic Area Forecast* chart in Canada. Although these products are experimental and not to be used as a sole source of icing information, they indicate percentages of icing probability and severity levels for a variety of altitudes at the time of the analysis or forecast, and can be used to increase your awareness of where icing conditions are likely.

In addition, since pilot reports (PIREPs) remain the only way of directly observing airframe icing in flight, it is important to check for the most recent icing PIREPs applicable to your route. To benefit your fellow pilots, it's also important to give PIREPs—especially if the actual conditions are not what were forecast. For the most part, the criteria for reporting icing intensity levels have remained the same over the years and are almost identical in both the United States and Canada: trace, light, moderate, and severe (see FAA *AIM* and TC *AIM*). The FAA's advisory circular on icing (AC 91-74) adds greater detail to the definitions and includes a fifth level—heavy.

The inherent subjectivity of the definitions means the same icing conditions could yield different reported intensity levels for different-sized aircraft. Therefore, icing PIREPs from aircraft similar to yours will provide a more accurate description of expected icing conditions—that is why aircraft *type* is also included in the reports. Keep in mind that a given icing intensity level reported from a larger airplane likely means significantly worse icing conditions for a smaller one.[13] The expanded intensity definitions from FAA AC 91-74 can be of help when interpreting an icing PIREP, or when giving one yourself:

> (1) Trace Icing. Ice becomes noticeable. The rate of accumulation is slightly greater than the rate of sublimation. A representative accretion rate for reference purposes is less than 1/4 inch (6 mm) per hour on the outer wing. The pilot should consider exiting the icing conditions before they become worse.

> (2) Light Icing. The rate of ice accumulation requires occasional cycling of manual deicing systems to minimize ice accretions on the airframe. A representative accretion rate for reference purposes is 1/4 inch to 1 inch (0.6 to 2.5 cm) per hour on the outer wing. The pilot should consider exiting the conditions.

(3) Moderate Icing. The rate of ice accumulation requires frequent cycling of manual deicing systems to minimize ice accretions on the airframe. A representative accretion rate for reference purposes is 1 to 3 inches (2.5 to 7.5 cm) per hour on the outer wing. The pilot should consider exiting the conditions as soon as possible.

(4) Heavy Icing. The rate of ice accumulation requires maximum use of the ice protection systems to minimize ice accretions on the airframe. A representative accretion rate for reference purposes is more than 3 inches (7.5 cm) per hour on the outer wing. Immediate exit from the conditions should be considered.

(5) Severe Icing. The rate of ice accumulation is such that ice protection systems fail to remove the accumulation of ice and ice accumulates in locations not normally prone to icing, such as areas aft of protected surfaces and areas identified by the manufacturer. Immediate exit from the condition is necessary.

(emphasis added)

To avoid the impression that flight into trace icing is acceptable—especially for aircraft not certified to fly into icing conditions—the FAA recently removed the phrase "it is not hazardous" from its definition of trace icing. Also note, for all intensity categories, it is recommended you consider exiting the icing conditions before they become worse (see underlined text [emphasis added] in definitions above). Regardless of the level of ice protection equipment aboard, if you experience **severe icing**, you must exit *immediately* since these systems will not be able to keep up with the rate of accumulation and remove the ice. Remember that the absence of reported or forecast icing doesn't mean you won't experience it; pilots have received PIREPs of "negative ice" but have found their aircraft covered in it anyway.

Never Take Off With Ice Contamination

Of the 583 airframe icing accidents included in the NTSB study cited earlier in this chapter, one in five occurred on takeoff. However, 70 percent of Part 121 icing accidents occurred during this critical phase of flight. When a pilot does something questionable with an airplane and suffers no serious consequences, it can often lead to repeat behavior. This is true for flight with airframe icing. There's no question that pilots have experienced airframe ice accretion in flight—even in aircraft not approved for flight into icing—and made it safely back to the ground; others have departed with ice, snow or frost on the wings and have made it safely into the air. Unfortunately, these successes tend to reinforce risky behavior on the part of pilots. The problem, of course, is that many pilots have encountered icing in flight and never made it safely to the ground, while others have attempted takeoffs with ice contamination and never made it safely into the air. For those pilots who've gotten away with it, we would ask, "How do you know when enough ice is too much? Do you really want to be a test pilot?"

This problem was highlighted during a historic investigation into the Air Ontario accident in March 1989, mentioned earlier in this chapter. The F-28 crashed just beyond the departure end of Runway 29 at Dryden, Ontario, after the crew attempted to takeoff with snow and ice on the wings. During their intermediate stop at Dryden, the crew members faced a dilemma: Company policy prohibited ground deicing procedures with one or more engines operating (for safety reasons). However, since Flight 1363 had been dispatched with an unserviceable auxiliary power unit (APU) and ground-start equipment was unavailable at Dryden, if they had shut the engines down, they and their passengers would have been stranded at Dryden. The pilots chose to keep one engine running. In hindsight, it would have been better if they had shut them both down and called it a night—especially for the 21 passengers, two pilots and flight attendant who did not survive the accident.

In its four-volume, 1,700-page report, the commission found that a somewhat permissive attitude toward airframe icing existed throughout the Canadian aviation community. After this accident, the entire Canadian aviation industry—airlines, regulators and pilots alike—learned a lesson: Tolerating even minute quantities of ice contaminants on aircraft-lifting surfaces can lead to tragedy. Three years later, after the crew of another F-28—USAir Flight 405—attempted a takeoff with ice-contaminated wings, the American aviation industry also learned its lesson. The NTSB discovered that during the 35 minutes since its last deicing, the airplane had accumulated ice and snow on its lifting surfaces; it stalled and crashed near the end of the runway, killing the captain, a flight attendant and 25 passengers (Report No: NTSB/AAR-93/02).

Regulations in both countries prohibit aircraft departures with frost, ice or snow adhering to the lifting or control surfaces. For almost 50 years, FAA regulations permitted pilots to depart with frost on the wings, stabilizing surfaces and control surfaces as long as it was *polished to make it smooth*. Not anymore. After at least 12 takeoff accidents since 1982 in which airplanes failed to generate enough lift despite the frost being polished smooth, and in response to a U.K. Air Accidents Investigation Branch (AAIB) recommendation to remove all references to "polished frost" from the FAA regulations due to a 2002 fatal takeoff accident in Birmingham, England, the FAA recently removed this provision from the Part 25, 91 and 135 icing regulations. It now requires complete removal of *all* frost from critical surfaces to regain *uncontaminated smoothness* prior to takeoff.[14]

Storing your aircraft in a heated hangar before flight is an effective way to protect it from icing. However, frozen contaminants can accumulate once it is brought outside into the elements. Even in clear weather, frost can form in active frost (ACT FRST) conditions where the temperature is at or below the frost point. Using wing and horizontal tail covers, waiting for the frost/ice to melt, or applying deicing fluid are other ways to attain a "clean" airframe. Deicing fluids are freezing-point depressants (FPD) that lower the freezing point of the water/ice on the aircraft's surface. Most also provide anti-icing protection, preventing contaminants from accumulating, but make sure you understand and adhere to the information provided in the **holdover time** tables (HOT tables) for the particular fluid you are using. Holdover time is the estimated time deicing or anti-icing fluid will prevent the formation of frost or ice, or the accumulation of snow, on the protected surfaces of an aircraft and is dependent on the prevailing weather conditions (e.g., ACT FRST, FZFG, RA, SN, FZDZ, FZRA, etc.).

The USAir Flight 405 accident at LaGuardia Airport might not have occurred if guidance provided in the HOT tables had been followed; according to investigators, the F-28 departed almost 25 minutes after the deice fluid's effective holdover time had expired. Though Type I (mostly propylene glycol) is the most common deicing fluid used, it easily flows off the aircraft and has shorter holdover times. Type II, III and IV fluids have thickening agents that allow for longer holdover times, but they cannot be used for airplanes with slow rotation speeds—the fluid may not flow off the wing prior to liftoff, causing a significant reduction in lift, or the remaining fluid/water residue might refreeze on takeoff or in flight.[15] Remember, only use deicing or anti-icing fluids in accordance with the manufacturer's instructions. If the flight or operating manual prohibits its use, you must find another way to deice your aircraft.

Avoid Icing Conditions in Flight

Whether you fly an aircraft that is approved or not approved for flight into icing, your best strategy is to avoid flying into icing conditions altogether. Knowing when and where airframe icing is most likely to occur is the first step in avoiding it; this is especially important if your aircraft is not certified for flight into icing conditions. Most airframe icing occurs above or near the freezing level in clouds or in freezing precipitation below clouds. Most is also confined to low clouds (e.g., stratus, ST; stratocumulus, SC; nimbostratus, NS) or mid-level clouds (e.g., altostratus, AS; altocumulus, AC). High clouds above 20,000 feet MSL (e.g., cirrus, CI; cirrostratus, CS; cirrocumulus, CC) are composed almost entirely of ice crystals and usually pose no serious airframe icing threat. Clouds of vertical development (e.g., cumulus, CU; towering cumulus, TCU; cumulonimbus, CB) begin in the low to mid-levels and can extend well into the upper regions of the troposphere and even penetrate the tropopause; depending on the latitude, CBs (or thunderstorm clouds) can reach altitudes as high as 50,000 to 60,000 feet, certainly higher than civil aircraft can fly. Because of strong lifting, clouds of vertical development contain large quantities of SLD and pose a serious icing threat regardless of altitude.

You can estimate the likelihood of airframe icing using your OAT gauge when flying in clouds. As a general rule, you can expect particular icing types depending on the general OAT ranges listed in Table 3-2.

OAT Range	Icing Type
0°C to -10°C	Clear
-10°C to -15°C	Mixed Clear & Rime
-15°C to -20°C	Rime

Table 3-2. Temperature ranges for ice formation.[16]

According to the FAA icing Advisory Circular 91-74, at temperatures below -20°C, most clouds are composed primarily of ice crystals and pose no significant icing hazard. However, be aware that the temperature ranges in Table 3-2 can drop further into the *minus* territory when you're flying in clouds of vertical development typical of unstable air.

Though airframe icing can occur whenever atmospheric conditions are right, the probability of an icing encounter is greatest during the six coldest months of the year: 81 percent of all airframe icing accidents cited in the NTSB study took place between the beginning of October and the end of March. The probability of encountering icing conditions—including SLD ice—also rises when flying near large bodies of water.[17] According to *A Pilot's Guide to In-Flight Icing,* a training course provided to pilots by NASA's GRC Icing Branch (see Helpful Resources at the end of this chapter), areas most likely to encounter icing conditions include the coastal regions of Alaska, British Columbia, Washington and Oregon, as well as a wide area that extends from the southern high plains, through the Great Lakes region, to the northeast coast of the United States and into the Canadian Maritime Provinces.[18] Keep in mind that icing accidents are not limited to these times of the year or geographical areas; they can occur wherever and whenever the meteorological conditions are favorable for ice formation.

Since updrafts are conducive to sustaining larger droplets, orographic lifting found on the windward side and crests of mountain ranges—such as the Cascades, Sierras, Rockies and Appalachians—can create severe icing conditions. Therefore, if your proposed flight plan takes you parallel to a mountain range on the windward side, consider altering it to fly on the leeward side to reduce your exposure.

Since no aircraft, including large transport category airplanes, are certified to continue flight in severe icing conditions, you must exit—or better still, avoid—SLD icing conditions. According to NASA's research*, SLD is usually found below 12,000 feet MSL in stable stratus clouds, but it can extend considerably higher in unstable cumulo-type clouds. Of course, you can also avoid serious SLD icing conditions underneath a cloud layer by steering clear of precipitation when the OAT is near or below freezing.

In summary, to reduce your chance of an icing encounter, avoid flight into clouds when the OAT is between 0°C and -20°C; don't fly into clouds of vertical development such as TCUs and CBs, or into precipitation when flying under a cloud deck when the OAT is near or below freezing; and avoid geographical locations that favor the formation of severe icing.

* All subsequent references to NASA research in this chapter can be found in the online course, *A Pilot's Guide to In-Flight Icing,* cited above. (See Helpful Resources at the end of this chapter.)

Exit Icing Conditions

Flight into icing conditions is sometimes unavoidable. If you fly an aircraft not certified for flight into icing and inadvertently encounter it, you must know how to immediately exit the icing environment before the situation worsens. If you fly an icing-certified aircraft and flight into icing conditions is unavoidable, you also need to know how to best escape from it, or at least minimize your exposure to it—especially if you encounter SLD and/or your IPS fails to remove the ice. No matter the aircraft type or flight operation, all pilots—even those who fly aircraft equipped with IPS—need to know how to exit severe icing conditions.

To escape from icing conditions your choices are up, down, sideways or some combination thereof. The million dollar question is, "Which is the best option?" The answer is, "It depends." In the belief that altitude is almost always a pilot's friend, you should consider climbing to get out of the icing environment—*provided your aircraft is capable of doing so.* That is the important caveat; your aircraft's service ceiling lowers and its climb performance diminishes the more ice is allowed to collect on its lifting surfaces. Therefore, climbing could actually make matters worse. Many pilots have tried to climb above an icing layer only to find their aircraft unable to make it, and they were then forced to descend back down with even more ice clinging to their aircraft than before. This is what happened to the Caravan pilot mentioned earlier in this chapter who barely made it to the runway after being forced to descend. He requested a climb to get above the icing layer, but here's what followed—in his words:

> I started my climb but noticed the aircraft would only climb 200 fpm at 120 KTS.
> I reached 7,500 ft. MSL and could no longer climb. The aircraft could hold altitude
> at maximum power and 120 KIAS for a short time. After approximately 10 minutes

the aircraft started to buffet at 120 KIAS and I reported severe clear ice. I pitched the aircraft nose down to keep from a stall. (ASRS Report No: 881246).

Exiting Icing in Cloud

If you are certain your aircraft has the performance capability to successfully climb to the desired altitude, then a climb may be your best option when flying in cloud: It could take you out of the cloud layer altogether, it could take you to a colder level (below -20°C) inside the cloud where it is mostly composed of ice crystals, or it could bring you into a temperature inversion with an OAT above 0°C. Your problems with ice aren't necessarily over, but at least you've made it to an altitude where it has stopped accumulating. Keep in mind that it will take considerable time for the aircraft to rid itself of the ice through the slow process of sublimation or the quicker process of melting (if flying in an above-freezing inversion)—especially if your aircraft is not equipped with icing protection.

If you are certain a descent can be made to an altitude that will either take you out of the cloud or to above-freezing temperatures, then a descent to exit icing conditions may also be a viable option—*provided you have adequate terrain clearance.* NASA research indicates that icing layers in stratus clouds, including SLD icing, seldom exceed a thickness of 3,000 feet; therefore, while climbing or descending, a change of altitude of only a few thousand feet may be all you need to exit the worst conditions. Remember that freezing levels rise in the summer, raising the altitudes in which icing may be encountered; therefore, it may not be possible to exit the icing conditions by climbing. Also, if your descent for landing will take you through icing conditions, it's advisable to remain above a cloud layer as long as possible before descending to minimize your exposure to ice accumulation.

While icing in stratiform clouds may extend hundreds of miles horizontally, its vertical extent may be limited; in cumuliform clouds it is usually significantly limited horizontally but may extend considerably higher vertically. According to NASA, most icing in stratiform clouds is found below 15,000 feet and in cumulus clouds below 27,000 feet (though it can be higher for both). Furthermore, even though the horizontal extent of a single cumulus cell averages between two and six miles—and greater for convective cloud lines or embedded (hidden) convective clouds in stratiform clouds—icing encounters in cumuliform clouds are likely to be short in duration but severe in intensity. Therefore, it is recommended to navigate laterally around these types of clouds. Also, as mentioned earlier, the most hazardous of cumuliform clouds—CB clouds—can extend to several thousand feet, making it impossible for even high-performance aircraft to exit the icing conditions by climbing. As you will discover later in this book, icing is not the only reason why you should completely avoid flying into or near CBs.

Exiting Icing in Precipitation

We mentioned that icing in precipitation (ICGIP) in the form of FZDZ or FZRA is typically associated with warm or stationary fronts in the winter and can be encountered while flying in a subfreezing layer below an above-freezing temperature inversion aloft. NASA research indicates that freezing precipitation, whether at the surface or at altitude, is typically less than 3,000 feet thick and is usually found below 12,000 feet, although its depth can be deeper and its extent higher. Don't think you are out of the woods if you encounter PL or SG, both of which do not adhere to the airframe; these are certain signs of a temperature inversion aloft with FZRA or FZDZ above and ahead of you. If you witness PL or SG falling while waiting on the ground, keep waiting until all the precipitation—including FZRA and FZDZ—has completely passed.

If the subfreezing layer extends to the surface, your best options for escape are limited to a climb into warmer air to melt the ice or a heading change to extricate your aircraft from the SLD icing. Again, the first option is achievable only if your aircraft has the ability to climb while it is accumulating severe ice and will make it to the warmer (above freezing) layer aloft. Knowing your aircraft's capabilities and as much real-time weather information as possible—from your preflight weather briefing, current PIREPs, ATC and any other available source—will assist you in this decision.

Your best choice, however, may be a 180-degree turn—especially if you are piloting an aircraft not equipped with icing protection or you are encountering FZDZ that has been produced through the collision-coalescence process. According to NASA's recent icing research, it is more common to find SLD in the form of FZDZ produced by this process in supercooled clouds than from freezing precipitation below temperature inversions. Therefore, you cannot always assume that a climb will take you to above-freezing temperatures. It might be best to exit the freezing precipitation laterally through a heading change. This is where your knowledge of the overall synoptic weather situation and specific conditions at your location is invaluable in helping you to discern the best course of action.

Finally, if you encounter icing in **frontal weather**—especially warm or stationary fronts in the winter—avoid flying parallel to the front. This will prolong your exposure to icing conditions that may extend for hundreds of miles. It is best to avoid flying near a front, but if you must, consider flying through the front—taking the shortest route perpendicular to it—and fly behind it rather than in front of it. An altitude change may be your only option to avoid the ice, but keep in mind that according to NASA research, icing layers associated with warm fronts may be 10,000 or more feet thick.

Minimize the Risk When Flying in Ice

If you fly an icing-certified aircraft and must occasionally fly in icing conditions, you must be able to do it as safely as possible. Not only does this involve effective use of your aircraft's IPS but also the ability to successfully monitor the icing's effect on the handling and performance of your aircraft.

Manage the Ice Protection Systems

Timely activation of the ice protection systems on your aircraft is the key to minimizing the detrimental effects of airframe icing. To prevent ice from forming on propellers, windshields, pitot tubes, static ports, and AOA, sideslip, and temperature sensors, many airline operating manuals recommend anti-ice heating systems be turned on *prior* to entering areas of visible moisture (clouds, precipitation, fog) when the OAT is 5°C or less. If you fly aircraft equipped with leading edge pneumatic deicing boots, unless the aircraft manufacturer specifies otherwise, the FAA and NTSB also recommend that you activate them *as soon as airframe icing is encountered*. Why? Because it's hazardous to wait. Several fatal accidents, including the Comair Embraer 120 accident discussed at the beginning of this chapter, involved pilots who waited to activate the pneumatic deice boots until the ice had accumulated to the manufacturer's minimum recommended thickness (typically 1/4 to 1/2 inch). In a 2008 *Safety Alert*, the NTSB points out that accidents like these, along with NASA laboratory and field icing research, confirm that ice accumulation as little as 1/4 inch thick can increase the stall speed by 25 to 40 knots and is sufficient to cause an airplane to depart from controlled flight when flying at normal approach speeds.[19] Therefore, since ice bridging is not the hazard it was once thought to be, and since pilots are often unable to accurately visually assess the thickness of ice on their wings (pilots involved in accidents and incidents have guessed wrong), the NTSB now recommends—unless otherwise dictated by the flight or operating manual—that pneumatic deice boots be activated as soon as airframe icing is encountered. As mentioned earlier, there will likely be residual inter-cycle ice that may not shed during the first inflation cycle, but it eventually will during subsequent cycles.

Manage the Automation

Most aircraft certified for flight in icing conditions are also equipped with **automatic flight control systems** (AFCS). Even a simple three-axis autopilot makes the job of flying easier—especially during high workload situations like the approach and landing. However, when flying in icing conditions, there are some automation "gotchas" that could literally kill you.

First, make it a practice to avoid using the flight director (FD) or autopilot in the vertical speed mode (V/S mode) during a climb. In an attempt to maintain the selected rate of climb, the airplane will sacrifice airspeed as ice accumulates and will stall if you are not paying attention. Most operators recommend using indicated airspeed mode (IAS mode) during the climb so that if ice goes undetected, rather than lose airspeed the autopilot will reduce the rate of climb to maintain the selected airspeed, thereby avoiding a stall.

Despite instructions in the Saab 340 airplane operating manual that an FD/autopilot climb with accreted ice on the airframe is only authorized in the IAS mode, the crew of American Eagle Flight 3008 used the V/S mode instead. What happened next was frightening for the three crew members and their 25 passengers. During the climb, the airplane picked up ice that went undetected by the crew, and in an attempt to maintain the selected vertical speed, the autopilot pitched the nose up, causing the airspeed to slow to 130 KIAS. Through 11,500 feet, and before the stall warning even activated, the crew lost control of the airplane in a roll upset. Fortunately, after a series of steep uncontrolled pitch and bank attitudes spanning 50 seconds, and after the FO began to manually operate the deice boots (the crew heard chunks of ice striking the airframe), the captain was able to regain control at 6,500 feet.[20]

Rather than using the autopilot, it may be better to hand-fly your aircraft when ice is present. Had the American Eagle pilot been hand-flying the Saab 340, he likely would have noticed the airspeed reduction, and he would have also physically felt the left rolling tendency the autopilot was countering while the airplane was approaching the stall. This caught the crew of Comair Flight 3272 off guard, as well as many other pilots involved in icing accidents and incidents. In a recent *Safety Alert for Operators* (SAFO), the FAA suggests that when ice is accumulating on the airplane, the autopilot should be disconnected at least once every five minutes to allow the crew to feel if any abnormal roll or pitch tendencies are being masked by the autopilot. The SAFO also warns that when you disconnect the autopilot, you should be prepared to correct for any "unusual control displacements" that could occur from ice-induced airflow disturbance.[21]

Don't Blindly Trust Your Aircraft's Stall Warning System

We mentioned in the previous chapter that two major goals of every pilot are to avoid hitting anything and avoid stalling. Nowhere is this latter goal truer than when flying in ice—the bottom line is to avoid loss of control from a stall. Since ice contamination reduces the stall AOA, proper stall warning is absent in aircraft equipped with warning systems that rely on the non-contaminated wing stall AOA. Many icing accidents have involved inadequate stall warning to the crew, including the Comair Flight 3272 accident mentioned at the beginning of this chapter. According to the report, the airplane departed from controlled flight *before* activation of the stick shaker. Some aircraft are designed with a **stall warning bias** that compensates for the reduced stall AOA by providing earlier warning at higher airspeeds when in icing conditions. However, you shouldn't completely trust this type of system either. A Circuit City Stores' Cessna Citation 560 was equipped with a modified system designed to provide earlier stall warning when flying in ice; however, it crashed while the crew was conducting an approach in icing conditions to Pueblo, Colorado, in part because the stall warning did not activate until after the stall occurred (Report No: NTSB/AAR-07/02).

You should also be aware that a stall with an ice-contaminated wing will likely be more sudden, and roll control more difficult, than a stall involving a clean wing. Most airplane wings are designed so the stall begins near the root and progresses toward the tips, creating a gradual stall that allows a degree of aileron control before the wing is fully stalled. However, the benefits of this desirable design feature can be completely lost with ice contamination since the smaller thin wingtips are more efficient ice collectors, and they are therefore subject to stalls at a lower-than-normal AOA. This can cause a very sudden deep stall in which roll control may be completely absent. According to FAA AC 91-74, should a stall or uncommanded roll occur in icing conditions, you should reduce the AOA and then add power, but expect to reduce the AOA more aggressively than normal and be willing to accept a greater loss of altitude than for a normal clean wing stall recovery.

It's important to note that tailplane stall recovery differs from wing stall recovery, and distinguishing between the two can be complex in theory and in practice. The consequences of a tailplane stall are similar to a wing stall—there is buffeting and the nose drops—but the recovery technique is quite different. In order to recover from a tailplane stall, the pilot must pull back on the elevator control, and if flaps have been extended, they should be raised. The use of power is likely to be aircraft-specific, but adding power is generally not recommended. In order to differentiate between a wing stall and a tailplane stall, remember that tailplane stalls can happen at higher airspeeds and are often associated with flap extension or increase in power. As mentioned earlier, tailplane stalls are relatively rare, so it is best to determine from the manufacturer if the airplane you are flying is subject to this phenomenon when flying in icing conditions. If it is, or you suspect it could be, consult the guidance provided in the manufacturer's flight or operating manual and FAA Advisory Circulars 91-74, *Pilot Guide: Flight in Icing Conditions*, and 23.143-1, *Ice Contaminated Tailplane Stall (ICTS)*.

Monitor, Monitor, Monitor!

The Cessna Citation accident in Colorado also highlighted another reason why airframe icing accidents occur: Pilots are simply not paying attention. In a study of 37 flight-crew-involved major air carrier accidents, the NTSB found that inadequate monitoring and/or error challenging by the non-flying pilot was involved in 31 (84 percent) of them![22] Crew-monitoring failures in aircraft accidents and incidents have risen to such prominence that the FAA revised its Advisory Circular 120-71, *Standard Operating Procedures* (SOP) *for Flight Deck Crewmembers*, by adding a special section devoted to crew monitoring and cross-checking (Appendix 19) and recommending that the use of the terms "pilot not flying" (PNF) be replaced with "**pilot monitoring**" (PM) to more accurately describe the vital role of the non-flying pilot as a monitor.

In spite of this, pilots are still failing to adequately monitor the state of their aircraft. Consider the Dash 8-400 accident mentioned at the start of this chapter: The aircraft's instrument indications—the IAS indicator, the low-speed cue, the downward-pointing airspeed trend vector, the changing colors of numbers on the IAS display, and the airplane's excessive nose-up pitch attitude—all accurately indicated the gradual loss of airspeed before the stick shaker activated, but these weren't detected by either of the two crew members of Colgan Air Flight 3407. According to the NTSB, the pilots had adequate time to respond to these cues but failed to do so because of a significant breakdown in their monitoring responsibilities, in part because of their failure to adhere to sterile cockpit procedures (Report No: NTSB/AAR-10/01). The sterile cockpit rule was discussed in the first two chapters, and the importance of complying with it will be emphasized again later in this book. Even though the rule applies to the critical phases of flight—all operations below 10,000 feet except cruise—you should consider *critical* any phase of flight where ice is accumulating on the airframe.

As these accidents demonstrate, it is essential that you effectively monitor your airspeed: You must fly at a faster indicated airspeed when flying in icing conditions to compensate for the reduction of lift. The question is how fast. The answer is you should maintain at least the minimum airspeed published in your flight or operating manual. Many manufacturers specify higher **landing reference airspeeds** when flying with ice-contamination ($V_{REF-ICE}$). If such information is unavailable, the FAA *Safety Alert for Operators* on inflight icing (SAFO 06016) recommends you fly at least 50 to 60 percent faster than the non-ice stall speed (V_S x 1.5 or 1.6). It furthermore recommends that you not accept an ATC-assigned airspeed lower than the above-mentioned speeds. It is vital to monitor not only airspeed but also engine performance. Be prepared to take action if your engines are losing power, or if higher power settings are required to maintain a given level of performance. In some icing accidents, the airspeed dropped from cruise to stall in less than three minutes!

Don't Wait Until It's Too Late

You should immediately exit icing conditions if one or more of the following cues are evident: diminishing airspeed, high power settings to maintain airspeed, unanticipated trim changes, or any evidence of SLD icing conditions. The main visual signs of the latter are ice accretions aft of where ice normally is found. According to the FAA AC 91-74 on icing, you should look for **aft accretions** on the upper or lower surface of the wing, beyond the active part of the deicing boots, on a propeller spinner (possibly extending to the blades) and on the unheated portions of side windows (possibly covering the entire window). Unusually extensive ice coverage and the presence of irregular jagged shapes such as ridges, fingers or feathers—especially in areas where ice usually doesn't form (e.g., nacelles, winglets, wing struts, etc.)—are also signs to look out for. The advisory circular warns to be especially alert when the OAT is below 5°C and you see rain or drizzle, or droplets that splash or splatter on impact.

You must immediately exit severe ice—no airplane is certified to fly in such conditions. Should you experience severe icing, disconnect the autopilot, hand-fly the airplane and immediately request emergency priority handling from ATC. This frees them up to provide the assistance you need. Consider the strategies discussed above for extricating yourself from these conditions while maintaining at least the minimum airspeed published in your flight or operating manual, or a speed that is at least 50 to 60 percent higher than the non-ice stall speed. If ATC issues a speed restriction that is too slow, tell the controller you're unable to comply, and maintain your airspeed. Be careful when deploying trailing edge flaps for the approach; if performance or handling characteristics suffer as a result, consider retracting them. The advisory circular on icing recommends gradually deploying them in stages, noting the aircraft's behavior during and after each deployment. Consult your flight or operating manual for use of flaps; many prohibit their use for extended periods in icing conditions. You will eventually need the landing gear down, but consider extending it later, rather than sooner, due to the possibility of excessive drag. If you need to maintain a higher speed all the way to the threshold, keep in mind you will need more runway distance for landing. If you are actively picking up ice on short final, there's a good chance you could be landing on an ice-contaminated runway, so be prepared for minimal braking capability. Finally, the advisory circular mentions several icing accidents that have occurred from stalling during the landing flare, so carry higher-than-normal power right to the touchdown.

The accident record is clear: Airframe icing continues to pose a threat to flight safety—even if you pilot a craft certified for flight into icing conditions. The best strategy is to avoid it altogether by maintaining a completely "clean" aircraft at all times—it's your *only* strategy if you fly an aircraft not certified for flight into icing conditions. Of course, ice avoidance is next to impossible if you are a professional pilot who regularly flies in so-called all-weather flight operations. However, you need to realize that every icing encounter is a new and unpredictable event, and just because you experienced no ill effects the last time you picked up ice, that doesn't mean you won't the next time.

Operating in icing conditions is no time to relax your guard. No matter how many times it takes to deice, never depart with an ice-contaminated aircraft. Make sure it is clean before takeoff. Use the knowledge obtained during your preflight weather briefing to avoid areas of ice. If you do encounter it in flight, don't expose yourself to it any longer than you have to; if at all possible, immediately exit the icing conditions, either vertically or horizontally—especially if you encounter severe ice. Make sure the IPS on your aircraft is working properly and use it accordingly. Don't trust the automation to keep you out of harm's way, and make sure you hand-fly the aircraft to feel for any unusual roll, pitch or trim movements. Be vigilant in monitoring the aircraft's performance, especially power-required and airspeed. Above all else, maintain a safe airspeed even if it means losing altitude. If it comes to that, divert to the nearest suitable airport and get you and your passengers safely on the ground as soon as possible.

Helpful Resources

Pilot Guide: Flight in Icing Conditions (FAA AC 91-74A). This advisory circular provides guidance to Part 91, 121, 125 and 135 fixed-wing aircraft operations regarding safe flight in icing conditions, what conditions should be avoided, and how to avoid or exit those conditions if they are encountered. (www.FAA.gov/regulations_policies/advisory_circulars/index.cfm/go/document.information/documentID/74471)

NASA's Glenn Research Center (GRC) Icing Branch provides two free online interactive courses: *A Pilot's Guide to Ground Icing* and *A Pilot's Guide to In-Flight Icing*. Excellent coverage of all aspects of airframe icing. (AircraftIcing.GRC.NASA.gov/courses.html)

AOPA Air Safety Institute free online interactive courses can increase your knowledge and help you avoid the threat of aircraft icing. (www.AOPA.org/education.aspx)
- *Weather Wise: Precipitation and Icing*
- *Accident Case Study: Airframe Icing*
- *Real Pilot Stories: Icing Encounter*

Bombardier Aerospace Customer Training provides a free eLearning course, *Icing Awareness: Preflight Considerations*. (www.BATraining.com/elearning/freecourses/icing)

Notes

1. K.R. Petty and C.D.J. Floyd, "A Statistical Review of Aviation Airframe Icing Accidents in the United States," *The 11th Conference on Aviation, Range, and Aerospace* (Hyannis, MA: October 4–8, 2004).

2. National Transportation Safety Board, *Testimony of Deborah A.P. Hersman, Chairman, National Transportation Safety Board, Before the Aviation Subcommittee, Committee on Transportation and Infrastructure, U.S. House of Representatives, Aircraft Icing* (Washington, DC: NTSB, February 24, 2010).

3. Petty and Floyd, "Statistical Review."

4. NTSB, *Testimony of Deborah A.P. Hersman.*

5. Federal Aviation Administration, *Pilot Guide: Flight in Icing Conditions*, AC 91-74A (Washington, DC: FAA, December 31, 2007).

6. Virgil P. Moshansky, *Commission of Inquiry Into the Air Ontario Crash at Dryden, Ontario: Final Report*, Volume I (Ottawa, ON: Minister of Supply & Services, Canada, 1992).

7. FAA, *Pilot Guide.*

8. John P. Dow Sr. and John Marwitz, "Rough Ice Is Bad Ice," *Aerosafety World* 4.12 (December 2009–January 2010): 32–37. Available online at FlightSafety.org/aerosafety-world-magazine/past-issues.

9. FAA, *Pilot Guide.*

10. National Transportation Safety Board, *Aircraft Accident Report: Crash During Approach and Landing, Circuit City Stores Inc., Cessna Citation 560, N500AT, Pueblo, Colorado, February 16, 2005*, NTSB/AAR-07/02 (Washington, DC: NTSB, January 23, 2007).

11. Dale Hiltner, Michael McKee, Karine LaNoé, and Gerald Gregorek, *DHC-6 Twin Otter Tailplane Airfoil Section Testing in the Ohio State University 7x10 Wind Tunnel*, NASA/CR-2000-209921/VOL 1 (Hanover, MD: NASA Glenn Research Center, September 2000).

12. Federal Aviation Administration, "Activation of Ice Protection, Final Rule," *Federal Register* 76.162 (Aug 22, 2011): 52241–52249. Available online at www.FederalRegister.gov/articles/2011/08/22/2011-21247/activation-of-ice-protection

13. Richard K. Jeck, *A History and Interpretation of Aircraft Icing Intensity Definitions and FAA Rules for Operating in Icing Conditions*, DOT/FAA/AR-01/91 (Washington, DC: Federal Aviation Administration, November 2001).

14. Federal Aviation Administration, "Removal of Regulations Allowing for Polished Frost, Final Rule," *Federal Register* 74.229 (December 1, 2009): 62691–62697. Available online at www.FederalRegister.gov/articles/2009/12/01/E9-28431/removal-of-regulations-allowing-for-polished-frost.

15. Paul Pellicano, "On Thin Ice: A Little Frost Won't Hurt…Or Will It?" *FAA Aviation News* 48.6 (November/December 2009): 6–8. Available online at www.FAA.gov/news/safety_briefing/2009.

16. Federal Aviation Administration, *Effect of Icing on Aircraft Control and Airplane Deice and Anti-Ice Systems*, AC 91-51A (Washington, DC: FAA, July 17, 1996).

17. Petty and Floyd, "Statistical Review."

18. National Aeronautics and Space Administration, Glenn Research Center Icing Branch, *A Pilot's Guide to In-Flight Icing* (online course), February 26, 2009. Available online at aircrafticing.grc.NASA.gov/courses.html.

19. National Transportation Safety Board, *NTSB Safety Alert*, SA-014 (Washington, DC: NTSB, December 2008). Available online at www.NTSB.gov/safety/safety_alerts.html.

20. National Transportation Safety Board, *Safety Recommendation*, A-06-48 through -51 (Washington, DC: NTSB, July 10, 2006).

21. Federal Aviation Administration, "In-Flight Icing, Turbo Propeller Powered Airplanes," *Safety Alert for Operators*, SAFO 06016 (Washington, DC: FAA, November 1, 2006).

22. National Transportation Safety Board, *A Review of Flightcrew-Involved, Major Accidents of U.S. Air Carriers, 1978 through 1990*, Safety Study NTSB/SS-94/01 (Washington, DC: NTSB, January 1994).

Pushing Weather

4

VFR FLIGHT INTO IMC

It must have been scary. Not only was the commercial pilot forced to land his Cessna 172 on a nearby private airstrip after encountering adverse weather while en route to his destination of Pagosa Springs, Colorado, but now he was facing the same situation again on this same trip—this time with his airplane low on fuel and no suitable airport within reach. According to the NTSB, the VFR pilot encountered IMC on this winter day and decided to conduct a precautionary landing on a snow-covered road. The road came to a "T," and the airplane slid off the end and crashed into several trees, causing substantial damage to the airplane. The NTSB determined the probable cause of this accident was the pilot's decision to initiate flight into known adverse weather, his delay in not taking remedial action by not making a precautionary landing sooner, and his continued **VFR flight into IMC** (NTSB Identification No: DEN06LA036).

The pilot and pilot-rated passenger in this accident were lucky—they both escaped injury and lived to talk about it. Unfortunately, most pilots aren't so lucky, as accidents resulting from VFR flight into IMC almost always kill. According to statistics compiled by the AOPA Air Safety Institute, 86 percent of all GA fixed-wing VFR-into-IMC accidents since 2002 were fatal. This is a staggering statistic given that the overall fatality rate—or **lethality index** as the Air Safety Institute calls it—for GA accidents was 20 percent or less.[1,2,3]

In these accidents, VFR pilots either depart into existing adverse weather or more typically continue VFR flight into gradually deteriorating weather, and while attempting to make it to their destination they inadvertently fly into IMC (defined as weather conditions below VFR weather minimums), losing their outside visual references. If they are VFR-only pilots, or pilots with inadequate instrument flying skills, they either fly under controlled flight into nearby terrain (CFIT) or experience spatial disorientation and lose control of their aircraft. The latter results in uncontrolled flight into terrain or in-flight structural failure due to the pilot overstressing the aircraft while recovering from an unusual attitude. (We will discuss the topic of **spatial disorientation** in greater detail in Chapter 9.) In fact, a study conducted by the University of Illinois in the 1950s found that pilots who lack sufficient instrument flying ability lose control of their airplane in an average of only 178 seconds once they lose outside visual references.[4]

The Accident Record

The NTSB and TSB conducted special safety studies into this problem and found that between the mid-1970s and 1980s, VFR flight into IMC was responsible for only approximately 5 percent of GA accidents, yet because of its high lethality index, it was responsible for 19 percent of all GA fatalities in the United States and a staggering 26 percent of all aviation fatalities in Canada.[5,6] Between the two countries, the VFR-into-IMC accident scenario played itself out about once every two days! The problem isn't restricted to North America: In the United Kingdom, for example, VFR flight into IMC accounted for 24 percent of all single-engine aircraft accidents over a 15-year period.[7] The situation has gradually improved since then—thanks in large part to increased education regarding the deadly nature of this threat—but it still persists. In the United States during the 1990s, a VFR-into-IMC accident occurred on average about once a week.[8] Even though it now repeats itself on average about twice a month, over the most recent 10-year period where statistics are available it has caused the loss of 518 lives in the United States.[9]

You may be under the impression that VFR-into-IMC accidents involve only relatively low-time inexperienced private pilots who do not possess an instrument rating. While they certainly are most prone to experience this type of accident, an AOPA Air Safety Institute study on GA weather-related accidents found more than a third of these types of accidents involved pilots with more than 1,000 flight hours, and one-quarter involved those with more than 2,000 hours.[10]

Commercial pilots aren't immune either. Alaska has the worst record for commercial aircraft accidents in the United States, including those resulting from attempted VFR flight into IMC. One study revealed that almost half of all Alaskan fatal air taxi accidents and two-thirds of fatal commuter accidents were the result of attempted VFR flight into IMC.[11] The National Institute for Occupational Safety and Health (NIOSH) also found that during the decade of the 1990s, airplane crashes were the leading cause of pilot occupational fatalities in that state—competing against logging and fishing—with "attempted VFR flight into IMC" having the dubious distinction of being responsible for the majority (63 percent) of them.[12]

A similar record exists in Canada. According to the TSB study, almost one-fifth of the VFR-into-IMC accidents involved pilots who had more than 3,000 hours, while commercial flight operations were responsible for more than a third of the accidents nationwide.[13] More than half of all commercial CFIT accidents in Canada over an 11-year period were caused by attempted VFR flight into IMC.[14] Finally, an extensive study of worldwide commercial CFIT accidents found about a fifth of them were caused by VFR flight into IMC.[15]

Though the accident record is slowly improving, one thing is abundantly clear: VFR flight into IMC is a still a significant cause of fatal commercial aircraft accidents and continues to be the number one cause of fatal GA weather-related accidents.

Environmental Factors

Several environmental factors increase the risk. The probability of experiencing a VFR-into-IMC accident obviously increases when attempting to fly VFR in less than VFR weather conditions, but also when flying at night and in mountainous terrain.

Adverse Weather

Adverse weather in the form of cloud, fog and precipitation that reduces ceiling and visibility values to below VMC is generally involved in these types of accidents. Minimum VFR cloud clearance and visibility values are established in the belief that pilots will have adequate outside visual references to safely maneuver their aircraft. However, some VFR-into-IMC accidents have occurred in weather conditions that were higher than basic VFR weather minimums, leading many to question the adequacy of these minimums.[16,17] Canada and the United States responded to this by modifying the minimums: During the 1990s, TC raised the traditional one mile visibility to two miles in uncontrolled airspace below 1,000 feet AGL in Canada, and both countries raised it to three miles at night.

Flying in the Dark

Accident statistics show that flying at night also increases the risk—by at least a factor of three. Even though only an estimated 10 percent of GA flight operations take place at night, 35 percent of VFR flight into IMC accidents in the United States, and 30 percent in Canada, occur after dark.[18,19,20] As you will find out in the discussion of night flying in Chapter 7, the accident risk significantly increases after dark, even in good VMC. How much more, then, is the risk in adverse weather at night?[21] As the VFR pilot in the following narrative from NASA's Aviation Safety Reporting System (ASRS) describes, it's simply harder to see inclement weather in the dark.

> I left XYZ later than I originally intended. In retrospect, I should not have left at a time that would require night VFR flight, given the cloud conditions… I did not notice entering IMC at first, and, in fact, remember being curious why the anti-collision lights were illuminating the cockpit and causing a strobe effect on the prop. Shortly thereafter I noticed that the turn coordinator was pegged in a left turn, the attitude indicator showed a 45-to-60-degree left bank, the directional gyro was spinning rapidly… I recognized the signs of an impending "graveyard spiral" and was able to return the plane to straight-and-level flight. The real cause of this incident was…lack of appreciation of the danger of night VFR.[22]

Flying in the Mountains

Statistics also make it clear that flying in the mountains significantly increases the risk of experiencing an aircraft accident, including one that results from attempted VFR flight into IMC.[23] A U.S. General Accounting Office (GAO) study of **mountain flying risks** for GA pilots found that compared to the other 37 contiguous states (lower 48), the probability of experiencing an aircraft accident increases by almost 40 percent when flying in the 11 western mountainous states and Hawaii. As you might suspect, the likelihood increases substantially when flying in Alaska—also designated a mountainous state—by 500 percent![24] The terrain, weather conditions and sparsity of weather reporting stations in Alaska are similar to what pilots flying in Northern Canada also must contend with. In a 10-year period in Canada, more than half of all VFR-into-IMC accidents occurred in mountainous terrain, while these areas only accounted for 22 percent of the total number of accidents. Of the 10 U.S. states with the highest percentage of these types of accidents over a similar time frame, eight are located in FAA-designated mountainous areas.[25, 26]

Moisture-laden Pacific air arriving over the mountains of the western United States and Canada is particularly troublesome for VFR pilots. When the ceiling drops below the mountain peaks, the options for a safe arrival are severely limited. Flying low under the cloud (a practice known as **scud running**) in the labyrinth of narrow-winding corridors below the mountaintops has trapped many pilots who were subsequently unable to escape. Pilots have flown into dead-end valleys where the terrain rises at a steeper angle than their aircraft could climb, and others have flown into narrow valleys and were unable to successfully perform a 180-degree turn. Many pilots have waited until it was too late to make that turn, causing them to collide with a mountainside; others were able to perform a successful

"180" only to find that the weather had closed in behind them. This happened to the pilot of a Cessna 172 while flying in mountainous terrain near Enumclaw, Washington. The ceiling began to lower while he was flying through a mountain pass, so he elected to turn back. However, after conducting a 180-degree turn he found the weather had boxed him in. He entered the clouds while conducting another 180 to avoid the weather, but when he popped out of the clouds it was too late—his airplane crashed into the side of a steep mountain. Luckily, the pilot escaped without serious injury (NTSB Identification No: SEA98LA070).

The Human Factor

Some VFR-into-IMC accidents occur after pilots knowingly depart into adverse weather. For example, a non-instrument-rated pilot lost control in flight after climbing into an overcast ceiling when departing Walker County Airport in Jasper, Alabama. Radar tapes indicate that before crashing about a mile west of the airport, the Beechcraft Baron changed from its intended southerly course to an S-shaped northerly track, then climbed and descended between 1,300 feet and 2,600 feet at speeds that varied between 100 knots and 160 knots. The NTSB concluded the probable cause of this accident was "the pilot's intentional visual flight into known instrument meteorological conditions, resulting in spatial disorientation and subsequent loss of control in flight." It was reported that the pilot had operated under VFR in IMC on several occasions prior to this accident (NTSB Identification No: NYC08FA231).

However, most VFR-into-IMC accidents occur after pilots depart in good (or at least acceptable) VFR weather conditions and subsequently encounter gradually deteriorating weather en route. This is what the pilot of a Piper PA-28 experienced on a VFR cross-country flight to Lancaster, California. The weather was VMC when he departed Fresno, but it started to deteriorate as the flight progressed. The surface weather chart indicated a cold front was moving through the area, and residents in the vicinity of the accident site reported that IMC prevailed that morning. The airline transport-certificated pilot (ATP) and his passenger were killed when their Cherokee struck mountainous terrain near Tehachapi, California. The probable cause was "the pilot's VFR flight into IMC and his ensuing failure to maintain terrain clearance" (NTSB Identification No: SEA07FA106).

No matter the scenario, the fact remains that the pilots involved made the decision, either by active deliberation or by passive default, to initiate or continue VFR flight into adverse weather. The million-dollar question is, "Why?" While investigators and researchers are still a long way from knowing the definitive answer(s) to this question, what they've learned so far may help you think about strategies to counter this hazard. The results of accident analyses and experimental studies seem to converge on three broad reasons why pilots might be tempted to fly VFR into IMC: They inaccurately perceive the nature of the threatening weather; they are subject to biases in their decision-making; and/or they feel pressure—either from themselves or others—to attempt VFR flight into IMC.

Inaccurate Assessment of Threatening Weather

You can end up "in the soup" if you fail to perceive the threatening nature of the weather, both during your preflight preparations and in the air while en route.

Assessing Weather Before Flight

For the 361 VFR-into-IMC accidents evaluated in its special study, the NTSB found that 42 percent of the pilot-related causes involved the way in which "weather information was obtained (or not obtained), assimilated, and used."[27] Also high on the list of causes were inadequacies in preflight planning and decision-making.

A prerequisite to an informed go/no-go decision is a proper preflight weather briefing. Sadly, too many pilots involved in these accidents either obtain an incomplete one or don't obtain one at all. For example, a non-instrument-rated pilot and passenger were killed when their airplane crashed into Lake Russell, near Lowndesville, South Carolina. Witnesses observed the Cessna 172 descending out of the "very low" overcast cloud layer (reported 100 to 200 feet above treetop level) in a 70-to 90-degree nose-down attitude before crashing into the lake. The accident report states the pilot did not obtain a weather briefing, and the probable cause was the pilot's "inadequate preflight planning and improper decision to continue flight into deteriorating weather conditions, which resulted in spatial disorientation after entering instrument flight conditions" (NTSB Identification No: ERA09LA527).

In a different accident, a collision with multiple treetops after encountering IMC at 500 feet AGL, shortly after departing Bellingham Airport in Washington State, was also blamed on inadequate preflight planning. The pilot of the Cessna 177 Cardinal, who survived the accident, later stated that he did not "receive a formal weather briefing from a flight service station" but relied solely "on the automated surface observation system report at the airport" (NTSB Identification No: SEA99LA025).

Finally—and regrettably—there are those who, though they receive an adequate preflight weather briefing and likely have an intellectual knowledge of the threatening nature of the weather information they obtain, deliberately choose to launch or continue flying into adverse weather anyway. Maybe they gauge the weather as less threatening than they should, or perhaps they believe they can handle it. A non-instrument-rated, 1,000-hour commercial pilot received a preflight weather briefing from the Great Falls Automated

Flight Service Station (AFSS) and two in-flight weather updates from Cedar City AFSS. He was advised all three times that "VFR flight was not recommended" and was warned by en route controllers at least twice of the deteriorating weather conditions ahead, but he still elected to continue on his cross-country flight. Unfortunately, he died after his Beechcraft Debonair crashed into high terrain about 10 NM north of Heber City, Utah (NTSB Identification No: SEA06FA036). (See the links in the Helpful Resources section for an AOPA case-study course re-creating the events of this tragic accident.)

Assessing Inflight Weather

In a study of 72 GA accidents that occurred between August 2003 and April 2004, the NTSB found that compared to shorter flights (50 NM or less), pilots on longer flights (more than 300 NM) were almost five times more likely to experience a weather-related (poor visibility/IMC) accident. One possible explanation is that weather conditions can change significantly on a long cross-country and pilots either fail to or are unable to obtain adequate weather updates for the remainder of their trip.[28]

Also, recent research indicates that even though the ability to accurately judge ceiling and visibility values from a moving airplane is a crucial component in avoiding these types of accidents, most pilots aren't very good at it. University of Illinois researchers recently had pilots fly a simulated VFR cross-country flight in weather conditions that had deteriorated to below-VFR minimums en route and found that the majority of them overestimated both visibility and ceiling values. They also discovered that those who diverted to avoid the simulated adverse weather were significantly more accurate in their estimates of flight visibility than those who continued to fly into it. They concluded that a major reason that pilots—especially those with fewer hours and less experience—continue flying into deteriorating weather is their inability to determine when they are in or nearing IMC.[29]

Decision-Making Errors and Biases

Faulty pilot decision-making is cited as a major cause or causal factor in VFR flight into IMC accidents. Research indicates that our decision-making abilities are sometimes less than optimal and can be subject to a variety of errors and biases.

Optimistic and Ability Biases

For example, most GA pilots believe they possess greater flying skill, are less likely to take risks in flight, and are less likely than their peers to experience an aircraft accident.[30] Research conducted by one of the authors of this book and his colleague also confirms that most VFR pilots believe they are less likely than others to experience a VFR-into-IMC accident and believe they are more capable than others at both avoiding and successfully flying out of IMC.[31] Pilots aren't the only ones who exhibit unrealistic optimism (**optimistic bias**) and an overestimation of their abilities (**ability bias**). For example, research indicates that most cigarette smokers see themselves at less risk of developing smoking-related health problems than the average smoker; most Americans believe they are more intelligent than their fellow citizens; and, comparing themselves to other professors, 94 percent of U.S. college professors believe they do above average work![32,33,34] And most automobile drivers not only believe they are more skillful than other drivers but also believe they are less likely to be involved in an automobile accident.[35]

Not all pilots can be above average, nor can all have a lower-than-average chance of experiencing an aircraft accident, yet that is what most pilots believe about themselves. Why is that? These two biases are part of a family of what are called **self-serving biases** that serve to protect an individual's ego by painting an unrealistic, positive self-view. In fact, the strength of these biases is significantly reduced in mildly depressed people and for those with lower self-esteem, and compared to so-called mentally healthy individuals, research indicates these people actually exhibit more accurate and realistic perceptions of reality![36] A subjective observation of most pilots would suggest they don't struggle with low self-esteem or depression; therefore, most pilots likely have a less accurate perception of reality. There is also considerable evidence supporting a link between a positive optimistic approach to life and reduced susceptibility to physical illnesses. The Catch-22 is the very biases that lead us to be overly optimistic regarding our chances of experiencing a VFR-into-IMC accident, and to be overconfident in our ability to avoid or successfully fly out of IMC, also seem to be crucial for good physical and mental health! These two biases correspond to the hazardous attitudes of invulnerability ("It won't happen to me") and macho ("I can do it") identified in the aeronautical decision-making (ADM) literature published by the FAA and TC. The irony is that even though these thought patterns might actually be good for a pilot's overall physical and mental well-being, they could indeed be *hazardous* when it comes to estimations of risk and their own flying ability.

Framing

How we **frame** our go/no-go decisions can also bias our choices. For example, if you had a choice between a sure win of $85 and an 85 percent chance to win $100, which would you choose? Most of us would likely avoid the risk and take the sure gain of $85. How about a choice between a sure loss of $85 and an 85 percent chance of losing $100? In this case, most of us would probably choose the risk of losing the $100. Research has found that when a problem is framed in terms of gains (as in the first example), we tend to be risk averse and choose the sure gain over the chance of a greater gain. But if framed in terms of losses (the second example), we tend to be risk seeking and choose the chance of a greater loss over the lesser sure loss. Simply framing identical information in two different ways—either positive or negative—is enough to bias decisions one way or the other. For example, when treatment for lung cancer is framed in terms of survival rates (*positive* frame) versus the corresponding mortality rates (*negative* frame), it changes patients' preference for one treatment over the other. We tend to avoid losses: As pioneer decision researcher Daniel Kahneman sums it up: "10 percent mortality is more frightening than 90 percent survival."[37]

There is also evidence that the way pilots frame their go/no-go flying decisions in the face of deteriorating weather influences their choices. For example, when faced with the decision of whether or not to continue flight into deteriorating weather, researchers found that pilots who framed their decision in terms of gains (certain gain of safely landing if they diverted versus a chance of a safe landing if they continued) were more likely to be risk averse and divert, while those who framed it in terms of losses (certain losses of diverting to an unplanned destination versus only a chance of an accident if they continued) were more likely to be risk seeking and continue the flight.[38]

Escalation Bias

Another bias occurs when a person feels they have too much invested to quit, even if the investment is in a failing course of action.[39] On one occasion, a student of one this book's authors was driving in the country late one night, and his gas gauge indicated near empty. He was in the middle of nowhere and didn't know if he would find a gas station before he ran out of fuel. If he turned back he felt confident he would make it to a station he had passed 10 minutes earlier, but he struggled with the decision because turning back would mean all of the resources (in terms of time and energy) that he had invested would be lost. (He probably also struggled with how to frame this problem: he knew the *certain loss* if he turned back, while if he kept going there was a chance he would make it and only a chance he would run out of gas.)

This entrapment, or **escalation bias**, may make it difficult for you to turn back in the face of poor weather and is one of many complex and usually unconscious psychological factors that influence decisions to press on into deteriorating weather—a condition commonly known as **get-home-itis**. In fact, researchers have discovered that unlike other types of accidents, those involving VFR flight into IMC tend to occur closer to the intended destination airport.[40] It appears that many of these accidents occur on the last leg of a return trip because the desire to get home overrides the ability to make a sound go/no-go decision. We have a name for it in aviation: the **last leg syndrome**. This commitment to continue gets stronger the closer you are to reaching your goal. In his book, provocatively titled *The Naked Pilot*, airline pilot-writer-psychologist David Beaty summarizes the problem by stating that a pilot, like an animal, "prefers to follow his nose, look at the goal and go where he is looking. The nearer the goal, the stronger the pull. The nearer the airport, the more hypnotic the drive to continue."[41]

Pressure, Pressure, Pressure

According to accident reports, a frequent causal factor in VFR-flight-into-IMC accidents is the pilot's self-induced pressure to complete the flight. This can occur for a variety of reasons, including the pressure to be somewhere on time. In one example, a non-instrument-rated private pilot attempted VFR flight into IMC and collided into mountainous terrain in Old Fort, North Carolina, killing himself and his two passengers. Once airborne, the pilot reported to a controller, "If the clouds come down a bit, I may want to do some scud-running." According to the accident report, he was trying to keep an important appointment (NTSB Identification No: ATL03FA061). In another tragic accident, the non-instrument-rated private pilot experienced spatial disorientation. Ten minutes after departing Santa Monica Municipal Airport, he lost control of his Beech Bonanza. The airplane stalled, entered a spin and crashed into a three-story apartment building in the Fairfax District of Los Angeles, California. The pilot, his three passengers, and one person in the apartment building were killed and seven others on the ground were seriously injured. The flight was destined for Sun Valley, Idaho, where the pilot was going to show property to two of his passengers. Before the flight, he and his passengers had been at the airport for at least eight hours waiting for weather conditions to clear. The accident investigators concluded that the pilot's "self-induced pressure to complete the flight" was a factor in the accident (NTSB Identification No: LAX03FA182).

Not only do pilots pressure themselves, but they feel pressure from others as well. Statistics show that compared to other GA accidents, a significantly higher proportion of VFR flight into IMC accidents carry passengers.[42] That means that the presence of others (e.g., friends, family, other pilots) can negatively influence a pilot's go/no-go decisions. Even others who are not aboard (e.g., employer, dispatch, ATC) can influence decision-making behavior. The accident record is replete with examples of how others' expectations have contributed to less-than-optimal decisions by pilots:

- A Delta Airlines B-727 is destroyed by impact forces and post-crash fire after takeoff from Dallas/Fort Worth International Airport, killing 12 passengers and two flight attendants. The pilots accepted an unexpected "taxi to position and hold" clearance from the controller even though they were number four with three airplanes ahead of them. According to the NTSB report, they taxied ahead of the other three aircraft, rushed their checklist, and 39 seconds later were cleared for takeoff. They failed to extend the flaps and slats for takeoff (Report No: NTSB/AAR-89/04).

- An entire college football team must use emergency exits and slides to evacuate a chartered McDonnell Douglas DC-9 that landed just off the side of the runway at Erie International Airport in Pennsylvania. Against his better judgment, the pilot knowingly continued an approach in questionable weather. The NTSB concluded the poor performance of the flight crew was partly due to the encouragement of the company dispatcher to attempt the approach, as well as the "subtle pressure provided by the business manager who occupied the jumpseat" (Report No: NTSB/AAR-86/02/SUM).

- A de Havilland DHC-6 Twin Otter crashes in bad weather on approach near Rockland, Maine, killing all but one of 18 people aboard. Again in this case, the pilot felt pressured to go against his better judgment. If he canceled the flight, he felt the company president would have harshly reprimanded him. The NTSB cited "inordinate management pressures" to complete the flight as factors in the accident (Report No: NTSB-AAR-80-5).

- A copilot acquiesces to his captain's failure to reject the takeoff—despite the fact that he knew that something wasn't right with the airplane's performance—and flies their ice-laden B-737 into the ice-covered Potomac River. All but five of the 79 aboard perish in the crash (Report No: NTSB-AAR-82-8).

These examples tell us what social psychologists have been telling us for years: Humans are profoundly influenced by others, even when we think we're not. Without us being aware of it, other people can influence us to the point that they are making decisions for us and in effect we are no longer flying the aircraft.

Avoiding a VFR-Flight-Into-IMC Accident

As you have seen, you can get trapped not only by the weather, but also by your own thinking. Fortunately, as with most other potential hazards to safe flight, there are a variety of strategies you can use to manage them. Adhering to the following suggestions will go a long way to help you avoid becoming a VFR-into-IMC accident statistic.

Always Keep Learning About Weather and Its Potential Threat to Safe Flight

You should never stop learning about weather and how it can affect your flight operations. Take a college-level meteorology or aviation weather course to learn all you can about meteorological processes. Then get in the habit of putting into practice what you've learned by making basic rule-of-thumb predictions based on existing weather conditions.

In essence, your aim is to become an amateur weather forecaster. As you gain experience it will become easier to recognize the signs of deteriorating weather, both through your personal observations when looking outside the cockpit and from a thorough understanding of aviation weather reports and forecasts. Pay particular attention to the characteristics of low-pressure systems and the signs of approaching warm fronts, both of which are associated with extensive areas of low ceilings and visibilities common to many of these types of accidents.

Adhere to Personal Weather Limits

Experienced VFR pilots know the value of establishing **personal weather minimums** and sticking to them even when the pressure to continue a flight becomes strong. Since VFR-into-IMC accidents have occurred in weather conditions that were higher than regulatory VFR weather minimums, it's crucial that you develop your own personal minimum ceiling and visibility values that are higher than basic minimums, and ensure that those are met before you depart or continue a flight. These should reflect your flight experience and comfort levels, and if you are a relatively new pilot, they most certainly should be conservative (i.e., considerably higher than regulatory minimums). In the words of one FAA official, personal weather minimums provide "a safety buffer between the demands of the situation and the extent of your skills."[43] Research supports their value as well. An extensive, multinational study of weather encounters found that pilots who reported flying into IMC had the most liberal personal weather minimums, while those who never flew into IMC had the most conservative ones.[44] See Helpful Resources at the end of this chapter for information designed to help you determine your own minimums and avoid inadvertent flight into IMC.

Once you've established your personal weather minimums, sticking to these minimums, even when under pressure to continue, is the key to avoiding inadvertent flight into IMC. Doing so prevents you from flying at low altitude below the clouds, which could put you dangerously close to terrain with few options to escape—a practice that is responsible for countless VFR-into-IMC accidents.

Get a Thorough Weather Briefing and Updates En Route

We already mentioned the sad fact that some pilots don't obtain a proper weather briefing before flight. In fact, the commercial pilot we mentioned earlier who conducted a precautionary landing to avoid flying into IMC and whose Cessna was damaged after sliding off a snow-covered road, wished he had obtained a weather briefing—but he consulted his friend for weather information instead! Always get a thorough weather briefing before every flight and keep updated en route. Not only will it help you to determine if the weather is suitable for a given flight, but the more briefings you obtain, the more knowledgeable you will become about various weather phenomena and their possible effects on flight. Obtaining weather advisories on a regular basis from FSS while en route will alert you to any significant changes in the weather ahead, especially ceiling and visibility values. Contact **Enroute Flight Advisory Service** (EFAS), also called Flight Watch, in the United States on 122.0 MHz below 18,000 feet (or assigned discrete frequency at or above FL180) or the appropriate Flight Information Centre (FIC) in Canada on the frequency listed in the *Canada Flight Supplement (CFS)*. If the weather is deteriorating, you need to start planning for possible alternates—the earlier the better.

On clear cool nights, regularly check nearby automatic terminal information service (ATIS) or automated surface observing system (ASOS) frequencies for the temperature/dew point (T/DP) spread to alert you of possible ground fog formation. Expect fog to form when the T/DP spread narrows to within 3°C. One of the authors of this book remembers a student commercial pilot candidate who returned to his home airport after completing a VFR night cross-country trip only to find that it was closed due to fog. He then flew back to the airport he had just returned from and found that it too had just closed because of fog. He really panicked after finding the same conditions at a third airport! Luckily, he was able to land at the nearby international airport before it too became socked in. He neglected to keep a close watch on the T/DP spread. The AOPA Air Safety Institute study found that 16 percent of GA weather-related accidents occur in fog with the majority of them involving fatalities.[45]

Be Careful in the Mountains

As discussed earlier, the accident record clearly demonstrates that mountains and weather do not mix. The old adage that says "You may be able to push the weather or push the mountains, but you can't push both" is true. To avoid getting trapped, your personal weather minimums must be higher in the mountains. Experienced mountain pilots in Canada have traditionally used "2,000 foot ceiling" and "5 miles visibility" as an absolute minimum. Are you an experienced mountain pilot? Both of the authors of this book consider themselves somewhat experienced. We've logged hundreds of hours of VFR flight in the mountains of western North America, and as instructors we've given mountain checks to dozens of pilots, yet we're not at all comfortable with "2,000 and 5" even over relatively flat terrain. Flight below a cloud deck in the mountains can be conducted relatively safely only if the ceiling and visibility are good and will remain so, plenty of altitude exists between your aircraft and the surface and/or obstructions (e.g., power transmission lines), you have sufficient room to complete at least a standard-rate 180-degree turn, and you've had plenty of experience in mountain flying techniques, including a thorough mountain check from an experienced mountain flying instructor.

Be Careful at Night

You've learned that the probability of inadvertent flight into IMC increases significantly when flying VFR at night. It's difficult to see worsening weather at night—even more so on *dark* nights when there is no moonlight or starlight. Not only should your nighttime personal VFR weather minimums be higher, but also you should fly over well-lit areas and at or above minimum terrain and obstacle clearance altitudes. If you are instrument-rated, fly at or above the minimum IFR altitude. (See Chapter 7 for a further discussion of risk-management strategies to use when flying at night.)

Obtain Supervised Exposure to Marginal Weather

Flight instructors usually don't fly in weather conditions they wouldn't allow their students to fly in because they want to set a good example by modeling the safest possible flying behavior. The downside of this is that the students only learn how to fly in good weather. But what happens when it turns bad? Sooner or later you will encounter unforeseen adverse weather and will need to know how to deal with it. The best way to prepare for this is to get some actual experience flying in less-than-ideal weather conditions with an experienced flight instructor. As long as you realize that flying in marginal weather is not the norm for VFR operations and that such training is only designed to help you effectively identify and escape from it should you inadvertently encounter it, there are instructors who would be glad to give you the benefit of their experience and provide you with such training. These instructors can help you learn the skills necessary to identify and keep your distance from adverse weather; accurately judge actual ceiling (in thousands of feet) and visibility values (in statute miles); determine minimum safe altitudes; and recognize and maintain a safe way out (which you should always have when flying in less-than-ideal weather). You'll also learn how to avoid overflying your visibility (similar to overdriving your headlights at night while driving) by slowing down and maneuvering at lower cruise speeds. Actual exposure to marginal VFR weather should also help you more fully appreciate the dangers of scud running, a practice that is almost always fatal at night or in the mountains. It's also a good time to learn the techniques necessary to conduct a safe **off-airport landing**. Remember, the goal of such training is to help you minimize your exposure to, and safely get out of, adverse weather, *not* teach you how to continue flying in it!

Get an Instrument Rating and Use It

A recent study found that pilots who do not hold an instrument rating are almost five times more likely than instrument-rated pilots to be involved in a weather-related accident.[46] If you have an instrument rating, you're less likely to experience a VFR-into-IMC accident because you're more likely to use your ticket in marginal VFR weather. Keeping instrument-current and filing IFR whenever the weather looks "iffy" for VFR flight is a sure way to avoid the hazards of scud running at low altitudes.

Don't Rely on VFR Flight in a Light Aircraft to Get Somewhere

If you have time to spare, take to the air.

No truer proverb has been spoken regarding VFR flight in a small aircraft. In spite of what the salesperson may have told you, VFR flight in a light aircraft is one of the *least* reliable forms of transportation. If you really must be somewhere, you should file IFR (weather, aircraft and pilot currency permitting) or choose another mode of transportation (e.g., automobile, bus, commercial air carrier). Flying—especially on long trips in light aircraft—carries with it enough pressures; why add to them by creating your own self-induced pressure to complete a trip? All VFR pilots should remind themselves of the maxim, "I don't have to *be* anywhere." Looking at VFR-into-IMC accident reports, it's sad to see how many pilots have pushed the weather in an attempt to make it on time to a funeral (an event most people feel they must attend) only to inadvertently make it to their own through inadvertent flight into IMC!

Remember: You Can Be Your Own Worst Enemy

Take to heart what you've learned in this chapter about the limits of your decision-making abilities. Sometimes your thinking is not always rational. In fact, knowing that your natural bent is to be unrealistically optimistic ("I won't ever suffer from a VFR-into-IMC accident") and overconfident in your own abilities ("I can avoid it or at least get out of it") is the most important step in stopping these biases from overruling your better judgment.

If you're faced with the decision to divert or continue VFR flight into marginal weather, resist the temptation to frame your options in terms of losses. When making mental risk assessments in this way, you know the losses involved in turning back include the certain loss to yourself and your passengers of unwanted taxi and overnight motel expenses, missed appointments, and generally unhappy people. Whereas you don't exactly know what the losses are if you continue—there's only a chance of a loss (an accident). You need to ask yourself: "How much is my life worth? Hotel expenses? Missed appointments? Some inconvenience?" Reframe your choice in terms of gains: "If I divert, I am certain to land safely, and if I continue the flight, there is only a chance I will make it through safely." So be careful how you frame your go/no-go decision—it could mean the difference between life and death!

Finally, don't forget the escalation bias—our propensity to continue on a course of action, even if it's a failing one, because we've invested too much to quit. Consider the scenario: The marginal VFR weather is below your personal weather minimums and as such you wouldn't normally fly in it. But you've been delayed several days due to weather and because you're so close to home—only 20 miles—you take what you perceive as an acceptable risk and fly in it anyway. (This decision parallels the fact that most long-shot bets at the racetrack—the most risky bets—take place during the last race of the day.) There are plenty of wonderful reasons for getting home—a return to your job, your family, and your own bed (with your spouse in it with you!). Almost without conscious awareness, you begin to reframe your situation to make it easier to continue home, but from the comfort of your chair as you're reading this chapter, you can clearly see how biased it is to narrow the choice down to only these two options: the *benefits* of arriving home and the *losses* of turning around and waiting. Instead, you need to think of the possible losses of continuing (an accident) and the benefits of diverting (staying alive). But that is not easy; it's been a long trip with many unplanned stops because of poor weather and you think to yourself, "My destination is only a short distance away. I've made it this far—I'm almost home!" You need to think of how many people have pushed the weather and didn't make it, and remember what you've learned in this chapter. Don't be the animal that prefers to follow his nose. Heed the words of W.C. Fields, who once said, "If at first you don't succeed try, try again. Then quit. No use being a damn fool about it."

Don't Let Someone Else Fly Your Aircraft

"Never fly in the same cockpit with someone braver than you." This is wise advice—but the person sitting next to you doesn't necessarily have to be braver. Remember the statistic quoted earlier: Your risk for a VFR-into-IMC accident rises significantly when you carry passengers. Do you remember the student who struggled with the decision to turn around to get gas for his car? Not only did the investment of time and energy contribute to his reluctance to turn back (escalation bias), but as it turns out, the presence of his girlfriend in the car did as well. He eventually did go back to purchase gas from that lonely station he'd passed earlier, but he said he struggled with the decision and it would have been much easier had she not been in the car with him.

Even if you're not consciously aware of it, the expectations of others—including those not in the cockpit with you—can negatively influence your go/no-go decisions to the point where they are making the decisions for you, and you in effect are no longer flying the aircraft. Therefore, you have to take stock of your situation and resist the pressure to let a family member, a friend, a boss or even a more experienced pilot-rated passenger fly your aircraft. As we pointed out in Chapter 1, sometimes it takes tact, diplomacy and assertiveness with passengers—and others—to ensure the flight goes the way it should. *You* are the pilot in command, and if something doesn't feel right, *you* have the full authority and responsibility to change that.

One more thing: The captain of the DC-9 that landed off the side of the runway at Erie Airport in Pennsylvania had been on duty 14.5 hours and awake for 19.5 hours at the time of the accident, which occurred just after 4:00 a.m.—a time when humans should usually be sleeping. Besides citing the subtle pressure from the college football team's business manager who occupied the jumpseat at the time, the NTSB also suggested that fatigue may have degraded the flight crew's ability to make sound decisions. When tired, you're more susceptible to pressure—from both yourself and others. Be aware that "fatigue makes cowards of us all" (a quote purportedly made popular by the famous U.S. football coach Vince Lombardi).

Consider a Precautionary Landing

If filing IFR is not an option and you find the weather is quickly closing in around you (likely because you delayed your go/no-go decision), consider conducting a precautionary off-airport landing. Though usually reserved for an aircraft systems emergency, this procedure has saved many pilots—including the one mentioned in the opening story of this chapter—from a weather emergency as well. An off-airport landing is not without risk—according to the AOPA Air Safety Institute, more than half involve no injuries and only 10 percent result in a fatality. But these consequences are much better than losing control of your aircraft or suffering a CFIT accident as a result of flying VFR into IMC.[47] If you don't know how to conduct an off-airport landing, then obtain training from an instructor who does. Apply what you've learned about the suitability of off-airport sites, and if safe to do so, land in a farmer's field. It could be your best option.

Don't Be Afraid to Ask for Help

Finally, don't wait until it is too late to ask ATC for help. Controllers can vector you around bad weather, give you minimum IFR obstacle clearance altitudes or provide you with an IFR clearance (assuming you're rated) if you elect to climb through the cloud to avoid a possible CFIT accident. Don't be more afraid of the possible legal repercussions than the hazardous weather itself. Yes, it's likely you will receive a follow-up call from the authorities after requiring priority assistance, but unless you're a repeat offender, or if it is clearly a case of willful, careless or reckless operation of the aircraft, a call is about all you will get. In the United States, the FAA's philosophy is that it would rather you learn from the situation and hopefully prevent a reoccurrence than take enforcement action. You need to ask which is worse: a talk with the authorities or a fatal accident.

Heeding the suggestions outlined in this chapter can help you avoid a VFR-into-IMC accident. Learning to recognize adverse weather before you get into it, and sticking to your personal weather minimums—especially at night or in mountainous terrain—will go a long way in reducing your risk. Obtaining an instrument rating and filing IFR in marginal weather provides extra insurance. But most importantly, the first step in making safer decisions is acknowledging that your decision-making is not always rational.

In addition, recognize that like most pilots, you're probably already biased to continue the flight rather than divert—just like most pilots seem doggedly determined to continue a bad approach rather than conduct a go-around. Recognizing this will help you to better counter the temptation to continue when the weather gets worse. If you find yourself uncomfortable about the weather you're flying in, trust your intuition. If it doesn't feel right, it probably isn't—no matter what someone else may think. If you get to the point where you need help, don't be afraid to ask for it. Better yet, *if in doubt, wait it out,* and ponder the insight attributed to Layton Bennett: "If you crash because of weather, you can be sure that your funeral will be held on a sunny day."

Helpful Resources

Do The Right Thing: Decision Making for Pilots is an online interactive course available from the AOPA Air Safety Institute that uses decision-making scenarios and other information to help you evaluate risk and determine your own personal weather minimums. The Institute offers several other free online interactive courses—including a re-creation and animation of the VFR flight into IMC accident near Heber, Utah, discussed in this chapter—that can increase your understanding of weather threats and improve your decision-making skills. (www.AOPA.org/education.aspx)

- *Weather Wise: VFR into IMC*
- *Accident Case Study: VFR Into IMC (accident in Heber, Utah)*
- *ASI Flight Risk Evaluator*
- *Weather Wise: Air Masses and Fronts*
- *Weather Wise: Ceiling and Visibility*
- *Mountain Flying*

The Art of Aeronautical Decision-Making is an online FAA Safety Team tutorial that draws upon the collective wisdom and expertise of VFR pilots and instructors who have learned from the mistakes of others. It provides invaluable information to help you determine your own minimums and avoid inadvertent flight into IMC. (www.FAAsafety.gov/gslac/ALC/course_content.aspx?pf=1&preview=true&cID=28)

Getting the Maximum from Personal Minimums is an excellent article about how to establish your own personal weather minimums. Published in the May/June 2006 edition of *FAA Aviation News*. (www.FAA.gov/news/safety_briefing/2006)

AOPA Air Safety Institute's *Safety Advisors* can assist you in understanding and reducing the threat of VFR flight into IMC. (www.AOPA.org/asf/publications/advisors.html)

- *Do The Right Thing: Decision Making for Pilots*
- *WeatherWise*
- *Spatial Disorientation*
- *Emergency Procedures*

Weather Decision-Making Guide is designed to help you—especially if you have relatively little weather-flying experience—develop skills in obtaining, interpreting and applying weather data to make safer weather flying decisions. (www.FAAsafety.gov/gslac/ALC/libview_normal.aspx?id=9724)

Notes

1. AOPA Air Safety Institute, *VFR Into IMC Accidents: 2012* (Frederick, MD). Available online at www.AOPA.org/asf/ntsb/vfrintoimc.cfm?window=3.

2. AOPA Air Safety Institute, *2008 Nall Report: General Aviation Accident Trends and Factors for 2007* (Frederick, MD: AOPA ASI, 2009).

3. AOPA Air Safety Institute, *2010 Nall Report: General Aviation Accident Trends and Factors for 2009* (Frederick, MD: AOPA ASI, 2011).

4. Leslie A. Bryan, Jesse W. Stonecipher and Karl Aron, "180-Degree Turn Experiment," *Aeronautics Bulletin* 52.11 (Urbana, IL: University of Illinois, 1954).

5. National Transportation Safety Board, *Safety Report: General Aviation Accidents Involving Visual Flight Rules Flight Into Instrument Meteorological Conditions*, NTSB/SR-89/01 (Washington, DC: NTSB, February 8, 1989).

6. Transportation Safety Board of Canada, *Report of a Safety Study on VFR Flight Into Adverse Weather*, Report No. 90-SP002 (Hull, Quebec: TSB, November 13, 1990).

7. David Learmount, "An Invasion of Privacy," *Flight International* 147 (April 12–18, 1995): 20–23.

8. Juliana Goh and Douglas Wiegmann, "Visual Flight Rules (VFR) Flight Into Instrument Meteorological Conditions (IMC): A Review of the Accident Data," *11th International Symposium on Aviation Psychology*, ed. R. Jensen (Columbus, OH: The Ohio State University, 2001).

9. AOPA, *2010 Nall Report*.

10. AOPA Air Safety Institute, *Safety Review, General Aviation Weather Accidents: An Analysis and Preventive Strategies* (Frederick, MD: AOPA ASI, 1996).

11. National Transportation Safety Board, *Safety Study: Aviation Safety in Alaska*, NTSB/SS-95/03 (Washington, DC: NTSB, 1995).

12. National Institute for Occupational Safety and Health, *Surveillance and Prevention of Occupational Injuries in Alaska: A Decade of Progress, 1990–1999* (Washington, DC: NIOSH, May 2002).

13. TSB, *Report of a Safety Study*.

14. Transportation Safety Board of Canada, *Aviation Occurrence Report: Controlled Flight Into Terrain, Western Straits Air, de Havilland DH3-3 (Turbine) Otter, C-FEBX, Campbell River, British Columbia 7NW, 27 September 1995*, Report No. A95H0012 (Hull, Quebec: TSB, 1996).

15. R. Khatwa and A.L.C. Roelen, "An Analysis of Controlled-Flight-Into-Terrain (CFIT) Accidents of Commercial Operators, 1988 Through 1994," *Flight Safety Digest* 15 (April–May 1996): 1–45.

16. TSB, *Report of a Safety Study*.

17. Federal Aviation Administration, "Night-Visual Flight Rules Visibility and Distance From Clouds Minimums: Final Rule," *Federal Register* 54.188 (September 29, 1989): 40324–40327.

18. NTSB, *Safety Report: General Aviation Accidents*.

19. TSB, *Report of a Safety Study*.

20. AOPA ASI, *Safety Review, General Aviation Weather Accidents*.

21. Dale R. Wilson, "Darkness Increases Risks of Flight," *Human Factors & Aviation Medicine* 46 (November–December 1999): 1–8. Available online at FlightSafety.org/archives-and-resources/publications/human-factors-aviation-medicine/human-factors-1999.

22. Office of the NASA Aviation Safety Reporting System, "Night for Day," *Callback* 199 (January 1996): 1.

23. Dale R. Wilson and Teresa A. Sloan, "VFR Flight Into IMC: Reducing the Hazard," *Journal of Aviation/Aerospace Education and Research* 13 (2003): 29-42.

24. United States General Accounting Office, *Aviation Safety: FAA Can Better Prepare General Aviation Pilots for Mountain Flying Risks*. Report to the Chairman, Subcommittee on Aviation, Committee on Commerce, Science and Transportation, U.S. Senate, Report No. GAO/RCED-94-15 (Washington, DC: U.S. GAO, December 1993).

25. NTSB, *Safety Report: General Aviation Accidents*.

26. TSB, *Report of a Safety Study*.

27. NTSB, *Safety Report: General Aviation Accidents*.

28. National Transportation Safety Board, *Risk Factors Associated With Weather-Related General Aviation Accidents*, Safety Study NTSB/SS-05/01 (Washington, DC: NTSB, September 7, 2005).

29. Douglas Wiegmann and Juliana Goh, *Visual Flight Rules (VFR) Flight Into Adverse Weather: An Empirical Investigation of Factors Affecting Pilot Decision Making*, FAA Technical Report ARL-00-15/FAA-00-8 (November 2000).

30. David O'Hare, "Pilots' Perception of Risks and Hazards in General Aviation," *Aviation, Space, and Environmental Medicine* 61 (July 1990): 599–603.

31. Dale R. Wilson and M. Fallshore, "Optimistic and Ability Biases in Pilots' Decisions and Perceptions of Risk Regarding VFR Flight Into IMC," *11th International Symposium on Aviation Psychology*, ed. R. Jensen (Columbus, OH: The Ohio State University, 2001).

32. F.P. McKenna, D.M. Warburton, and M. Winwood, "Exploring the Limits of Optimism: The Case of Smokers' Decision Making," *British Journal of Psychology* 84 (1993): 389–394.

33. Ruth C. Wylie, *The Self-Concept Vol. 2: Theory and Research on Selected Topics* (Lincoln, NE: University of Nebraska Press, 1979).

34. K. Patricia Cross, "Not *Can* But *Will* College Teaching Be Improved?" *Renewing and Evaluating Teaching: New Directions for Higher Education* 17, ed. John A. Centra (Spring 1977): 1–15.

35. Ola Svenson, "Are We All Less Risky and More Skillful Than Our Fellow Drivers?" *Acta Psychologica* 47 (February 1981): 143–148.

36. Shelley E. Taylor and Jonathon D. Brown, "Illusion and Well-Being: A Social Psychological Perspective on Mental Health," *Psychological Bulletin* 103 (March 1988): 193–210.

37. Daniel Kahneman, "Judgment and Decision Making: A Personal View," *Psychological Science* 2 (May 1991): 142–145.

38. David O'Hare and Tracy Smitheram, "'Pressing on' Into Deteriorating Weather Conditions: An Application of Behavioral Decision Theory to Pilot Decision Making," *The International Journal of Aviation Psychology* 5 (1995): 351–370.

39. Allan I. Teger, *Too Much Invested to Quit* (New York: Pergamon Press, 1980).

40. David O'Hare and Douglas A. Wiegmann, *Continued VFR Flight Into IMC: Situational Awareness or Risky Decision Making?* Final Report (April 17, 2003).

41. David Beaty, *The Naked Pilot: The Human Factor in Aircraft Accidents* (Shrewsbury, England: Airlife Publishing, 1995).

42. Goh and Wiegmann, "Visual Flight Rules".

43. Susan Parson, "Getting the Maximum from Personal Minimums," *FAA Aviation News* 45 (May/June 2006): 1–8.

44. David R. Hunter, Monica Martinussen, Mark Wiggins, and David O'Hare, *VFR Into IMC—Who, What, When, Where, and Why Summary of Results from the International Weather Encounters Study.* Available online at www.AVHF.com.

45. AOPA ASI, *Safety Review, General Aviation Weather Accidents.*

46. NTSB, *Risk Factors.*

47. AOPA Air Safety Institute, *Accident Case Study: VFR Into IMC* (online course). Available online at www.AOPA.org/Education/Accident-Case-Studies.aspx.

Can You Survive the Ride?

5

LOW-LEVEL WIND SHEAR

"Tower, Delta one ninety-one heavy, out here in the rain, feels good." Delta Air Lines Flight 191, a Lockheed L-1011 TriStar on final approach for Runway 17L, had just been handed off to the Dallas/Fort Worth (DFW) tower controller, and the rain may have provided a welcome respite for Captain Edward Connors as he spoke those words on that hot muggy afternoon in August. What he didn't know was that less than two minutes later, those same rain showers would kill him, his flight crew, five cabin attendants, 126 passengers, and the driver of an automobile who happened to be in the wrong place at the wrong time.

They knew they were about to fly through a thunderstorm—the FO pointed out the lightning coming from the cloud ahead—but they had no idea how deadly it was. How could they? Not only were there no convective SIGMETs, severe weather warnings or center weather advisories in effect at the time, but when the cell showed up just north of the airport, ATC called it a "little rain shower." The pilots of a Learjet and a Boeing 727 on the same approach just ahead of the L-1011 later testified that the cell "looked harmless"—both aircraft flew through the heavy rain shower only moments before the TriStar and experienced no ill effects.

Post-crash analysis determined that Delta Flight 191 flew into a cell that in only eight minutes had rapidly intensified from a weak (level 1) to a very strong (level 4) thunderstorm by the time of the accident. During the last 38 seconds of the flight, the L-1011 encountered a sudden headwind of 26 knots that changed into a tailwind component of 46 knots, creating a total horizontal wind shear of 72 knots; it also flew into downdrafts that reached almost 3,000 feet per minute. Seven seconds before impact the captain called for a go-around, but it was too late—at 280 feet above the ground and still descending at almost 5,000 feet per minute, the jumbo jet ran out of room to recover and struck terrain about a mile short of the runway. The airplane slid across a highway—hitting a car—and broke apart after crashing into two water tanks at the south end of the airport (Report No: NTSB/AAR-86/05).

Flight 191 was brought down by severe **wind shear** caused by one of nature's most deadly sources of it—a **microburst**. Wind shear is defined as a change in wind speed and/or direction over a very short distance either horizontally, vertically, or both. Changes in wind velocity are normally gradual, occurring over large distances and producing little or no effect on a given flight. However, a significant change in wind speed or direction over a very short distance can dramatically affect an aircraft's performance. As an aviator you know that an airplane is supported in flight by the dynamic reactions it makes with its wings as it flies through the air, and it doesn't matter at what speed the air mass itself might be moving over the earth's surface. As a practical matter, we combine the speed and direction of the air mass—wind velocity—with the aircraft's airspeed to determine ground speed. However, the creation of lift is a function of *airspeed*. Therefore, if the velocity of the air in which the airplane is traveling abruptly changes—say within hundreds of feet, or within a few seconds—because of the aircraft's inertia, its airspeed (and hence its lift) will also abruptly change, even if only for a few seconds while the airplane accelerates or decelerates back to its original airspeed.

The Effects of Low-Level Wind Shear

Wind shear can either increase or decrease aircraft performance. Sometimes, as in the case of Delta Flight 191, it does both. An **increasing performance shear** (also called an increasing headwind shear or decreasing tailwind shear) occurs when flying from a calm wind into a headwind or from a tailwind into a calm wind or headwind. This results in a sudden increase in airspeed and lift, causing the airplane to initially climb above its altitude or glide path while passing through the shear. A more severe hazard is presented by a **decreasing performance shear** (also called an increasing tailwind shear or decreasing headwind shear), which occurs when flying from a calm wind into a tailwind or from a headwind into a calm wind or a tailwind. This results in a sudden decrease in airspeed and lift, causing the airplane to initially descend below its altitude or glide path while passing through the shear.

Wind shear can occur at any altitude, and other than often producing turbulence—which at times can be substantial—it usually poses no major threat when encountered during the cruise portion of flight, since there is usually plenty of altitude and airspeed to recover. However, wind shear can be a recipe for disaster if encountered at low speeds when flying close to the ground, such as during takeoff, initial climb, and the approach and landing phases of flight. The airplane is in a low energy state both in terms of airspeed (kinetic energy) and altitude (potential energy), leaving little room to avoid a stall or premature contact with the ground. It's during these encounters with **low-level wind shear** (LLWS)—defined as wind shear below 2,000 feet AGL—when flights are most vulnerable. This is especially true with severe wind shear that causes an indicated airspeed change of more than 15 knots or a change in vertical speed of more than 500 feet per minute.[1]

LLWS was responsible for almost 40 percent of airline accident fatalities in the United States in the decade between 1976 and 1986,[2] and according to the FAA Advisory Circular (AC) 00-54, *Pilot Windshear Guide*, between 1964 and 1986 it led to at least 32 major accidents and incidents that claimed more than 600 lives.[3] Worldwide, ICAO calculates at least 1,400 lives have been lost since 1943 from wind shear-related accidents.[4] More recently, a Flight Safety Foundation task force found that almost one-third of worldwide approach-and-landing accidents and serious incidents involving jet and turboprop airplanes weighing more than 12,500 pounds MCTOW were due to adverse wind conditions including strong crosswinds, tail winds, and *wind shear*.[5] According to recent statistics compiled by Boeing, during the first decade of this new millennium almost 100 people died in accidents involving wind shear/thunderstorms in commercial jet aircraft with MCTOWs greater than 60,000 pounds.[6]

The aviation industry learned its lessons from Delta Flight 191 and from several major accidents leading up to it. In response to a fatal wind shear accident involving an Eastern Air Lines Boeing 727 that crashed short of the runway while conducting an approach at John F. Kennedy Airport in 1975, the FAA installed **low-level wind shear alert systems** (LLWAS) at major airports throughout the country. Consisting of strategically located wind velocity sensors surrounding the runway complex, an LLWAS is designed to alert air traffic controllers, who in turn can warn pilots, of the presence of wind shear at or near the airport. Since then, the number of LLWAS has considerably increased and their accuracy has significantly improved, especially since the advent of terminal Doppler weather radar (TDWR) in the 1990s.

In the early 1990s the FAA also implemented regulations requiring all turbine-powered commercial aircraft operating under Part 121 to be equipped with onboard **wind shear detection systems**. These provide pilots with real-time wind shear warning and flight guidance and are of two different varieties. Airborne warning and flight guidance systems provide reactive wind shear (RWS) warnings by alerting crews to the presence of wind shear and providing escape guidance information *after* entering wind shear. Airborne detection and avoidance systems provide predictive wind shear (PWS) warnings by alerting crews to the presence of wind shear and providing flight guidance information *before* encountering it. This technology has also improved over the years with aircraft manufacturers using a variety of systems such as forward-looking Doppler radar (which measures the frequency difference between the outbound and return signals—the Doppler shift) and **enhanced ground proximity warning systems** (EGPWS) to provide these warnings. Finally, in response to the Delta Flight 191 accident, the FAA also issued new regulations requiring commercial air carrier flight crews to receive ground and simulator training in LLWS recognition, avoidance and recovery.

The adoption of these new technologies and training requirements have paid off in terms of safety: The last fatal commercial airliner accident in the United States due to wind shear occurred in 1994 when a USAir DC-9 crashed during a missed approach at Charlotte/Douglas International Airport, killing 37 passengers (Report No: NTSB/AAR-95/03). The NTSB's accident database reveals that wind shear was a causal factor in only three air carrier accidents since the USAir accident—thankfully, none involved fatalities. However, that same search turned up approximately 70 wind shear-related accidents (mostly light GA aircraft) since the Charlotte crash with at least six of them involving fatalities. Even though the frequency of accidents has diminished over the years—especially in scheduled air carrier operations—LLWS still remains a formidable hazard that has to be guarded against by both airline and GA pilots alike.

Sources of Wind Shear

In order to avoid an encounter with severe LLWS, you need to understand the atmospheric and environmental conditions that lead to its formation. In order to escape from an encounter with LLWS, you need to understand its characteristics. There are generally two types of wind shear: **convective** and **non-convective**. Convective wind shear involves vertical convective currents that often produce clouds such as TCUs and CBs. Though relatively short-lived, this type of shear has been responsible for many fatal accidents. We will focus specifically on this deadly hazard after first outlining the major types of non-convective wind shear that you might encounter. These include wind shear produced by topographical effects, temperature inversions and low-level nocturnal jets, fronts, and sea breezes.

Topographical Effects

One way in which wind velocity near the earth's surface is significantly altered within a short distance is when mechanical obstructions interfere with its flow. Natural obstructions such as mountains, hills and tall trees, or human-made barriers such as buildings or hangars located at or near an airport, can cause not only wind gusts but also wind shear for arriving and departing aircraft. Besides altering its direction, the friction caused by terrain and obstructions retards the wind speed closest to the surface, producing rotating currents of air known as eddies, with characteristic downdrafts on the downwind side and updrafts on the upwind side. Often accompanied by turbulence—called **mechanical turbulence**— these eddies can cause rapid fluctuations in airspeed. The greater the surface roughness and the faster the wind speed, the greater the mechanical turbulence. Many pilots experience wind shear on approach to small airports surrounded by tall trees. As the aircraft descends below the level of the trees, the headwind abruptly decreases, leading to a sudden reduction in airspeed. The aircraft nose drops and the pilot must add power in order to maintain the approach angle and maintain airspeed.

Pilots who regularly fly in and out of airports located in mountainous terrain can testify to a variety of wind shear hazards. Downsloping winds flowing down narrow valleys tend to speed up from a funnel effect, creating wind shear at their interface with the prevailing winds. Cool air (from ice- or snow-covered slopes or plateaus) under the influence of gravity can also drain into low-lying regions, producing winds in otherwise calm areas.

Mountains can also interact with the prevailing wind to produce weather phenomena known as **mountain waves**. Like undulating waves of water that form downstream of a rock obstructing the flow, a similar phenomenon occurs in the atmosphere when air is forced up the windward side of a mountain range and then down on its leeward side. If the layer of air is stable and a sufficient wind is blowing—at least 20 knots at mountaintop level, but more common at higher speeds—a series of waves can develop that can extend as high as the tropopause and up to hundreds of miles downwind. When sufficient moisture is present, cap clouds that hug the mountaintops, and unique formations of standing lenticular (lens-shaped) clouds that form at the crests of the waves, can be seen. Common lenticular clouds include stratocumulus standing lenticular (SCSL), altocumulus standing lenticular (ACSL), and cirrocumulus standing lenticular (CCSL). These remain relatively stationary over the ground even though the wind is rapidly flowing through them: They form on the updraft as the air expands and cools, and dissipate on the downdraft when the air is compressed and heated. A cloud type particularly associated with extreme turbulence and wind shear is the rotor cloud (or roll cloud). It forms below the wave crests as the flow interacts with the surface boundary layer. At their greatest intensity—usually closest to the mountains— 180-degree wind reversals can cause significant wind shear and extreme turbulence when flying in or below these clouds.

A final hazard when flying in mountain wave conditions is the presence of severe down-drafts on the lee side of the mountains, especially near the mountain peaks. Exceeding 5,000 feet per minute—certainly greater than the climb capability of most aircraft, even the high-performance variety—these downdrafts are another reason to know how to safely navigate around mountain wave conditions.

Temperature Inversions and Low-Level Nocturnal Jets

There is typically a difference in velocity between winds at the surface and those aloft. Since surface friction retards wind speed and changes its direction, when climbing through the first 2,000 to 4,000 feet above the earth's surface—the area known as the boundary layer or friction layer—wind speed usually increases and its direction veers to the right in the Northern Hemisphere due to Coriolis force. The difference is usually gradual and causes no appreciable change in aircraft performance. However, under certain atmospheric conditions a more abrupt change in wind velocity can occur, causing significant wind shear. One such condition involves a temperature inversion.

In the daytime, wind speed at the top of the surface boundary layer is reduced due to mixing with the slower more turbulent air below. However, during the development of a nighttime surface-based temperature inversion, the cooler stable air below acts like a wedge that inhibits this mixing. As a result, the surface wind speed decreases and the winds at the top of the boundary layer become detached and speed up as they ride on top of it, creating what is known as a **low-level nocturnal jet**. Sometimes the wind speed exceeds 50 knots— hence the name "jet," after the fast-moving upper level jet streams found at the tropopause. Downsloping winds from mountain ranges can enhance this effect. The speed of the jet increases as the evening progresses and reaches its maximum when the vertical temperature gradient is the greatest. The temperature gradient reverses, and the inversion—and hence the jet—is broken up after daybreak when the sun heats up the earth's surface. A low-level jet can form just about anywhere if the conditions are right, but the phenomenon is most common in the central plains of the United States and prairies of Canada.

Fronts

The meeting place of two air masses that possess distinctly different temperature and moisture characteristics is called a front. Since the wind moves at different directions and speeds in these separate air masses—especially closer to the surface—any wind shear that may be encountered will be concentrated within this narrow transition zone. Frontal slopes are quite shallow (typically less than 2 degrees), so whether flying near a surface cold front (cold air advancing) or a warm front (warm air advancing) you're more likely to fly through this zone during a climb after a takeoff or when conducting an approach.

The greater the temperature difference between the two air masses and the greater the speed of the front, the higher the probability of significant wind shear. Therefore, wind shear is more likely to be encountered with faster-moving cold fronts. Significant frontal activity in North America is at its highest in the colder months of fall, winter and spring, averaging four to five systems per month for the central and northeast portions of the United States and about one per month for the southern states and areas east of the Rockies.[7] The presence of a semi-permanent low-pressure system near the Aleutian Islands and off the coast of Alaska (the Aleutian Low) also means frequent frontal onslaughts—especially in the winter—over Alaska, the Yukon, British Columbia, and the Pacific Northwest. Not all fronts produce hazardous wind shear, but you need to exercise caution when flying in their vicinity; the FAA's *Pilot Windshear Guide* reports that 14 percent of wind shear-related accidents and incidents occurring between 1959 and 1983 were the result of frontal activity.

Sea Breeze Fronts

A more localized frontal system known to cause wind shear is a sea breeze front. **Sea breezes** occur during the day near large bodies of water when the ground heats up faster than the adjacent water. Common on relatively clear calm summer days, a circulation develops in the afternoon when the cooler air over the water flows toward the lower pressure created by the rising warm air over the land. Though not as severe or extensive as air mass frontal activity, the leading edge of this onshore flow—the front—can create sudden wind shear for pilots operating at airports located on the coastline. As the temperature gradient changes at night a flow reversal occurs, causing an offshore flow toward the water. However, for a variety of reasons, the nighttime **land breeze** is usually less intense than the daytime sea breeze.

Convective Wind Shear

Convective wind shear results from vertical motions produced by updrafts and downdrafts occurring in an unstable atmosphere. Visible signs of this motion include clouds of vertical development—CU, TCU, CB, and altocumulus castellanus (ACCAS). Significant wind shear is possible whenever convective activity takes place, but the most severe is associated with mature CB clouds, otherwise known as **thunderstorms**. These convective storms produce almost every type of weather hazard known to pilots—turbulence, hail, icing, tornadoes, downbursts—including severe vertical and horizontal wind shears. It's no wonder that two out of three wind shear accidents/incidents studied by the FAA between 1959 and 1983 were the result of convective activity.[8]

Three ingredients are necessary for the formation of a thunderstorm: high moisture content, unstable air to high altitudes, and a lifting agent (e.g., fronts, mountains, convective heating). Thunderstorms also progress through three stages: cumulus, mature and dissipating. Updrafts dominate the cumulus stage with significant amounts of latent heat released into the atmosphere as the moisture condenses. A CU cloud rapidly grows into a TCU and eventually into a full-blown CB that can sometimes extend up to the tropopause or higher (50,000 to 60,000 feet). As precipitation-sized droplets grow they are held aloft in the upper portion of the cloud by the updrafts. Eventually, the growing weight of the precipitation overcomes the updraft, and it begins to fall rapidly to earth. Updrafts in that part of the cell begin to reverse as precipitation drags the air down with it, while evaporative cooling of the falling droplets further increases the strength of the downdraft. This marks the beginning of the most intense stage of the storm's development—the mature stage—in which updrafts and downdrafts coexist, sometimes for hours depending on the structure and type of the thunderstorm. The system eventually enters the dissipating stage as the precipitation and downdrafts take over the storm, cutting off its energy source.

It's during the mature and dissipating stages that severe wind shear is most likely. Downdrafts within a cell can reach high speeds, creating a downburst that spreads out laterally in all directions near the surface—the outburst. The leading edge of this fast-moving cold air—the gust front—acts very much like a rapid-moving cold front pushing warm air aloft and producing significant turbulence and strong wind shear for aircraft flying through it.

Even more deadly is what sometimes lurks inside the downburst—a microburst. This small, yet tremendously intense downdraft produces dangerous wind shears for aircraft. According to the FAA *AIM*, downdrafts within a microburst are typically less than a mile in diameter until reaching about 1,000 to 3,000 feet AGL where they spread out horizontally and extend to usually no more than 2½ miles in diameter. They seldom last longer than 15 minutes after they strike the ground and no more than five minutes at their greatest intensity (although they may continue for up to an hour if several of them are concentrated in a line). However, even though microbursts occur on a very small scale in both space and time, they can produce downdrafts of up to 6,000 feet per minute and wind speeds near the surface as high as 150 knots—an extremely dangerous situation for any aircraft.

But it gets worse. When flying through a microburst, the first encounter with the gust front acts like a sucker punch, luring a pilot into taking actions that may actually make matters worse. Here's the scenario: You're flying an ILS approach maintaining a 3-degree glide slope when at approximately 1,000 feet AGL you experience a sudden increase in airspeed (increasing performance shear)—that's the gust front. This increasing performance shear brings you well above the glide path, so you reduce power—close to idle—to bring your aircraft back down to it. As you re-intercept the glide path your airplane unexpectedly begins to rapidly descend—that's the downdraft. You increase thrust to arrest this descent, but because you're flying a turbojet, the engines take several seconds to spool up. Then the last punch hits you: a sudden decrease in airspeed (decreasing performance shear)—that's the tailwind shear. You lower the nose to avoid a stall, but it's too late. You run out of airspeed, altitude, and ideas and fly into the terrain. This is what happened to the crew of Delta Flight 191 and many pilots before and after them. Figure 5-1 illustrates the problem.

The chance of experiencing LLWS produced by a microburst is very low: According to the FAA's *Pilot Windshear Guide*, it is estimated that only 5 percent of thunderstorms produce them. However, that is no reason to be complacent. As many pilots who have gone before you have discovered, it only takes one encounter with nature's deadliest form of wind shear to bring an aircraft down.

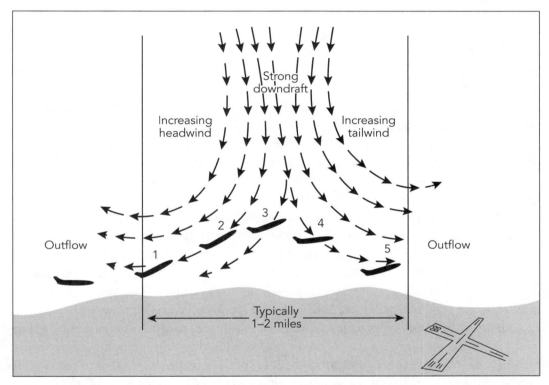

Figure 5-1. Flight through a microburst: 1 and 2, increasing headwind; 3 and 4, downdraft; 4 and 5, tailwind.[9]

Managing the Hazard of Low-Level Wind Shear

Whatever its source, wind shear poses a serious threat to safe flight operations—especially severe wind shear during the takeoff, approach or landing phases of flight. As with most weather-related hazards, the best strategy is to avoid flight into areas where an LLWS encounter is likely. However, this is not always possible. Therefore, you must also be able to recognize the in-flight indications of wind shear and know the procedures necessary to safely free yourself from its grasp.

Know Before You Go

You can ascertain the likelihood of LLWS before you even launch. If wind shear has been reported or is forecast, it will be included in the standard weather briefing relayed to you by the flight service specialist during your preflight weather briefing. If you are starting out with a self-briefing using the Internet or other sources, look for evidence of weather phenomena that are known to produce wind shear: gusty winds; mechanical turbulence; significant mountain winds, including mountain wave activity; nocturnal temperature inversions; frontal activity, especially fast-moving cold fronts; daytime sea breezes; and especially, convective activity.

LLWS observed at or near an airport will be included in routine (METAR) and special (SPECI) aviation weather reports. Also look for CB and TCU cloud types and the descriptor "TS" (thunderstorm) for indications of possible LLWS. A PIREP is a valuable source of real-time information about the location and severity of LLWS—FSS will even issue an Urgent PIREP if LLWS is reported—but keep in mind that, like a METAR, a PIREP indicates observed (not forecast) conditions and its value diminishes with time.

Obtaining forecasts of expected wind shear conditions for the time period in which your flight will take place is a critical component of gathering your preflight weather information. One of the best products for providing the "big picture" synopsis of the overall weather is the area forecast (FA). Use it in conjunction with AIRMET, SIGMET, and convective SIGMET weather advisories to determine forecast en route weather and to interpolate expected conditions at airports that do not have a terminal aerodrome forecast (TAF). If non-convective LLWS is expected it will be spelled out in the AIRMET and in the main body of the FA. Also, any forecast of convective activity implies the possibility of LLWS. The issuance of a severe weather watch for your area of flight is, of course, immediate cause for concern. These define areas of possible severe thunderstorms or tornadoes. A severe thunderstorm is one that produces hail at least three-quarter inch in diameter and/or winds of at least 50 knots.

The following is an example of a non-convective LLWS forecast from a TAF:

13012KT. . .WS020/27055KT.

In this example, LLWS is forecast from the surface to 2,000 feet AGL—not just *at* 2,000 feet—with surface winds from 130 degrees at 12 knots shearing to 270 degrees at 55 knots at 2,000 feet AGL.[10] Again, TS or other convective activity mentioned in a TAF also implies LLWS.

Finally, be aware that turbulence and wind shear often go together; therefore, suspect wind shear whenever moderate or greater turbulence is forecast.

Avoid Wind Shear by Recognizing the Signs

The best way to manage the hazard of LLWS is to avoid it altogether. A major reason for doing so is because some encounters can't be successfully recovered from. Even the best and most expeditious recovery procedure may not be enough to overcome the airspeed loss or downdrafts associated with certain types of wind shear—especially from microbursts. Therefore, the best way to avoid it is to recognize the signs of wind-shear-producing conditions. Whether the information you obtain is from weather reports or forecasts, from onboard instruments and avionics, or from simply looking outside the cockpit, when in flight you should always be on the lookout for weather conditions conducive to wind shear.

Non-Convective LLWS

Look for significant differences in surface winds and the winds aloft. For example, if the surface wind is light or calm when departing or approaching an airport surrounded by tall trees, you could experience a loss of airspeed and performance once above the tree line on takeoff or below it on an approach.

Watch for signs of mountain wave activity. These include high wind speeds at mountaintop level and the presence of standing lenticular, cap or rotor clouds. Keep in mind, however, that no clouds will be present if there is insufficient moisture. Suspect rotor activity—the location of the most severe turbulence and wind shear—if there is smoke or dust rotating below the crests of lenticular clouds. To avoid the most hazardous turbulence, wind shear and downdrafts, fly well above the mountain peaks—by several thousand feet. Flying too close to mountainous terrain leaves little room to maneuver if you encounter strong turbulence or downdrafts. In fact, the NTSB cited "downdrafts" as the most common causal factor (53 percent) of 384 GA turbulence-related accidents that occurred over a 10-year period between 1992 and 2001.[11] Flying in mountainous terrain involves an elevated level of risk—especially for the untrained—so take a mountain flying course from an experienced instructor if you are contemplating flight in or near the mountains. The instructor will be able to show you the techniques needed to minimize these risks, such as crossing mountaintops at sufficient altitude, crossing mountain ridges at a 45-degree angle for a quicker escape if you experience a downdraft, and slowing to maneuvering speed (V_A) to minimize the effects of turbulence.

Be on guard for LLWS if you suspect the presence of a surface-based temperature inversion. Fog, smoke and haze will often form below the top of an inversion. The conditions that favor its formation include a clear sky, cool night, and little or no surface wind. Be especially cautious if the winds on top of the inversion differ substantially from those at the surface. You may experience LLWS when climbing or descending through the layer. According to FAA Advisory Circular 00-6A, *Aviation Weather*, you should expect a wind-shear zone in a temperature inversion whenever the wind speed between 2,000 and 4,000 feet AGL is at least 25 knots.

Flying through a frontal zone is usually inadvisable due to a variety of weather hazards, not the least of which is wind shear. However, some frontal systems are more benign than others. If this is the case, and you are considering penetrating a front, keep in mind that the odds of encountering wind shear increase with the speed of the front and the temperature difference between the two air masses. An FAA Accident Prevention Program Bulletin (*Wind Shear,* FAA-P-8740-40) advises you to expect wind shear if the front is moving at least 30 knots and the temperature difference across the front is at least 5°C.

Convective LLWS

Minimizing your chances of a wind shear encounter from a microburst is just one of many reasons to avoid flying in or near a thunderstorm. (Other reasons include airframe icing, discussed in Chapter 3, and turbulence, hail and tornadoes, to name a few). In fact, it is recommended that you steer clear of them by at least 20 NM. However, the chance of avoiding them is less if your flight is at night, or if you encounter embedded CBs while flying in IMC. Be prepared for microburst wind shear whenever you are flying close to the ground in the presence of convective activity. You are most likely to encounter it in heavy rain showers beneath a thunderstorm (a **wet microburst**). But be on your guard for strong downdrafts even when there is no precipitation. **Dry microbursts** often form below thunderstorms that have high bases; the rapid downward movement of air is caused by strong cooling as the precipitation evaporates on the way down. Visible signs of a dry microburst include virga—visible streaks or wisps of precipitation falling just below the cloud base but evaporating before reaching the ground—and blowing dust or a ring of dust at the surface below the virga. Strong radar returns (or "echoes") on your onboard weather radar indicate heavy precipitation and possible microburst wind shear. However, weaker returns do not mean its absence; dry microbursts often paint weak echoes due to the lack of any appreciable precipitation. Convective clouds, turbulence and a large temperature/dew point (T/DP) spread also indicate the possibility of dry microburst conditions.

The FAA *Pilot Windshear Guide* (AC 00-54) lists important indications of an increased likelihood of microburst wind shear. These include:

- The presence of convective weather near your intended flight path (including localized strong winds, blowing dust, rings of dust and tornado-like features; heavy precipitation; rain showers, lightning, virga, moderate or greater turbulence; and a 17- to 28-degree Celsius T/DP spread);

- A PIREP of an airspeed gain or loss of 15 knots or more;

- A wind shear warning from onboard wind shear detection equipment (from your aircraft or a report from another aircraft); and

- An LLWAS alert or wind velocity change (especially 20 knots or greater).

The *Pilot Windshear Guide* also states that if any of the above are present—especially within a few miles of the runway along your intended flight path for takeoff or landing, below 1,000 feet AGL—you should consider delaying your departure or approach. After all, how long does a typical microburst last? A 10- to 20-minute delay may be all that is necessary to avoid the threat.

Once again, the best way to manage the LLWS hazard is avoidance. However, having done all you can to avoid it, there's still no guarantee you won't encounter it—sometimes flight into wind shear is unavoidable. Airline pilots, for example, must contend with convective weather on a regular basis—especially during the spring and summer. Using onboard weather radar, ATC assistance, and their own vision to detect it, pilots do their best to deviate around convective activity, but sometimes are unable to. In fact, an MIT study found that two-thirds of airline pilots who encountered convective weather while approaching the DFW Airport chose—for seemingly legitimate reasons and otherwise—to penetrate the cells rather than deviate from them. The MIT researchers found the likelihood of pilots penetrating heavy, and even extreme, thunderstorm cells (level 3, 4 and 5 intensity levels) increased the closer the aircraft was to the airport (within 15 miles), the further behind schedule they were, and when other aircraft ahead of them had successfully penetrated the cells.[12] Unfortunately, the MIT study did not address the possible reasons why pilots chose to penetrate intense convective storms rather than divert. However, if one of the reasons airline pilots continue an approach in the presence of convective weather is because of reduced fuel margins—perhaps because of weather delays that cut into fuel reserves—then it makes sense to carry extra contingency fuel if there is even the slightest possibility of delay due to cell activity. Avoid getting yourself into a situation where your only two choices are flight through a thunderstorm or running out of fuel.

Take Precautions if Flying in Suspected Wind Shear Conditions

Sometimes atmospheric conditions don't clearly indicate that you should delay your departure or approach, but the conditions still make you suspicious of the possibility of wind shear. If you elect to fly in such conditions, the *Pilot Windshear Guide* (FAA AC 00-54) recommends you take precautionary measures to minimize the effects of wind shear if you inadvertently encounter it. The following discussion highlights some precautions mentioned in AC 00-54 that you might want to consider when taking off or landing. These were developed primarily for pilots flying commercial jet transport airplanes and assume a possible encounter with the most severe form of wind shear—a microburst. However, some of these precautions and principles apply to other types of flight operations and, unless otherwise dictated by your aircraft's flight or operating manual or company SOPs, may be of benefit to you regardless of the type of aircraft you fly. Keep in mind that these precautions should in no way be considered an endorsement to fly in such conditions, and don't think that using them will immunize you from a wind shear accident.

Precautions During Takeoff and Climb

Prior to an LLWS encounter, it is ideal for the aircraft to be in as high an energy state as possible; this means faster airspeed and/or higher altitude. Of course, this conflicts with the requirements of takeoff and landing—slower airspeed and low altitude. But there are ways to compensate for this and increase the likelihood of a successful outcome should you inadvertently encounter wind shear close to the ground.

You obviously need to consider wind direction and speed when choosing a runway for departure, but if at all possible, choose a runway with a departure path that takes you away from the suspected wind shear and has the longest length. The latter will provide more room to stop if you need to perform a rejected takeoff (RTO) because of a wind shear encounter.

Full-powered takeoffs are the norm in light aircraft, but in larger aircraft, reduced-power takeoffs are often used. If you suspect wind shear, use the maximum rated power (or "thrust," a term usually used for jet aircraft) for takeoff. Not only will this shorten the takeoff roll and provide the best climb performance after liftoff, but if a wind shear is subsequently encountered, you will have more altitude in which to recover and somewhat less workload during the recovery because the recommended maximum thrust will already be set.

A higher-than-normal rotation speed will also put the aircraft in a higher energy state, but consult the *Pilot Windshear Guide* regarding this technique—it contains a number of caveats and cautions regarding higher rotation speeds for large aircraft. For small aircraft, use a normal rotation speed, ensuring there is an adequate buffer above stall speed in case you encounter wind shear and turbulence shortly after liftoff. Allow the airspeed to build to the maximum rate-of-climb speed (V_Y) as per the AFM, and fly that speed until reaching a safe altitude.

If you are using an FD for pitch guidance after liftoff, do not use it in the speed-referenced mode because it might command an inappropriate pitch attitude if you encounter wind shear. During a wind shear recovery procedure you want to maintain a specific attitude, because airspeed will likely fluctuate. The FD on Delta Flight 191 was speed-referenced, so when it was placed in the takeoff go-around (TOGA) mode it did not command the optimum pitch attitude for wind shear recovery. It instead gave pitch commands to capture a specific airspeed. Unless your aircraft is equipped with an FD that provides wind shear recovery guidance, you should ignore the FD's pitch commands, use the pitch attitude recommended for a wind shear escape procedure (to be discussed later), and disengage it if time permits.

Precautions During Approach to Landing

Appropriate runway selection in suspected wind shear conditions is also important for the approach and landing. If the surface winds are amenable, choose the longest runway that keeps you away from suspected areas of wind shear. If possible, also choose a runway with approach slope guidance such as an ILS, visual approach slope indicator (VASI), or precision approach path indicator (PAPI) system. This will allow you to quickly detect any deviations from the desired approach path.

The approach should be stabilized within specified altitude, airspeed and vertical speed parameters as early as possible so that you can quickly detect any changes in them. For transport category airplanes, this usually means the aircraft should be configured for landing (gear and flaps) and all checklists and call-outs accomplished before reaching a 1,000-foot height above the runway. Chapter 10 describes the stabilized approach concept in more detail.

If you observe an increase in airspeed and deviation above the glide slope, don't immediately reduce power. You might be experiencing the first part of a microburst, and downdrafts and a tailwind shear could follow. By delaying a response, you will have more energy in terms of airspeed and altitude, and the engines will be adequately spooled up and ready for a possible maximum thrust climb (especially important for turbojet aircraft). Keep in mind that if a tailwind shear doesn't materialize, you may have to conduct a go-around if you are unable to regain the desired glide path and suitable airspeed prior to landing.

If you are piloting a large transport category airplane, FAA AC 00-54 suggests you consider using a slightly higher approach speed in likely LLWS conditions. This will increase climb capability and provide a greater margin above the stall speed in case you need to recover from a wind shear encounter. The trade-off, however, is an increase in landing roll. For example, a 20 knot speed increase may translate into a 25 percent increase in stopping distance. Therefore, the FAA AC 00-54 recommends that pilots of large aircraft increase speed by no more than 20 knots above V_{REF} (normal approach speed for given landing weight). However, before using this technique you should consider how this extra landing distance, along with other landing performance factors (e.g., winds; runway length, slope and surface condition; density altitude; tire and brake condition; aircraft weight; thrust reverser availability; etc.), will affect your aircraft's ability to stop within a safe distance.

Regardless of the type of aircraft you fly, remember to factor the wind gusts in to your landing speed. When taking off or landing in gusty wind conditions (not necessarily microburst conditions), it is recommended you increase your airspeed by half the gust factor. For example, if you are landing on Runway 15 and the wind is from 150 degrees at 15 knots, gusting to 25 knots, the gust factor is 10 knots. In this case, if your normal approach speed is 80 KIAS, you should increase it by 5 knots to 85 KIAS.

If so equipped, use the autopilot, FD and autothrottle during the approach. These autoflight systems often make the job of conducting a stabilized approach easier by reducing workload and providing more time for you to monitor the instruments and the outside environment. However, these systems may mask the onset of wind shear and may not perform well in gusty winds and turbulence. Be ready to turn them off and hand-fly the airplane at any point during the approach.

Flaps should be selected to the minimum setting required for both takeoff and approach to allow for the maximum climb rate after takeoff or during a wind shear escape procedure on the approach. However, this suggestion is highly type-specific, so consult the AFM or company SOPs.

Monitor Aircraft Performance

By its very nature, wind shear can form quickly and, as the accident record clearly indicates, can completely catch pilots off guard. Therefore, a key component in quickly recognizing the onset of wind shear is a thorough knowledge of your aircraft's normal performance parameters. Knowing how your aircraft performs under normal conditions—especially vertical flight path performance—will help you to better recognize when it isn't. For example, you should know typically how fast your airplane accelerates to rotation speed on takeoff so you are better able to recognize when something isn't right. You also should be intimately familiar with how the initial takeoff and climb pitch attitudes look (both on the gauges and with reference to the outside horizon) along with the corresponding rate of climb and airspeeds on the climb-out leg.

The same applies to the approach: Know the normal descent performance parameters for a given airspeed and configuration including pitch attitude, power setting, and rate of descent. You should also know what the full-power go-around pitch attitude looks like, since this is usually the minimum pitch attitude recommended for a wind shear recovery procedure.

In addition to knowing your aircraft's normal performance characteristics, it is equally important that you are able to effectively monitor any change in them during the critical portions of flight (i.e., takeoff, climb, approach and landing). In addition to following some of the suggestions listed earlier (e.g., using auto-flight systems and flying a stabilized approach), you can increase your effectiveness by being vigilant in monitoring any deviations indicated by the vertical flight path instruments, such as the vertical speed indicator, airspeed indicator, altimeter, and glide slope displacements on the ILS. If you fly in a crewed flight deck, the pilot monitoring (PM) should carefully monitor these parameters and verbally call out any deviations from them. Of course, if you've read the previous chapters, you also understand the importance of complying with the sterile cockpit rule in maintaining effective monitoring. It's not only good practice for maintaining vigilance in avoiding collisions on the ground and in the air, and for monitoring aircraft performance parameters when flying in ice, but also when flying in suspected wind shear conditions.

Know How to Recover From a Wind Shear Encounter

You should initiate a **wind shear recovery procedure** if you encounter LLWS or suspect an encounter is imminent. Indications of an LLWS encounter include, but are not limited to, airspeed stagnation or fluctuation on the takeoff roll, airspeed loss or gain on the approach, updrafts or downdrafts, or large pitch and/or trim changes required to maintain the desired flight path. FAA AC 00-54 recommends that pilots be prepared to execute an immediate recovery if any deviations occur in excess of the criteria shown in Table 5-1.

Takeoff & Approach	± 15 KIAS
	± 500 fpm vertical speed
	± 5 degrees pitch
Approach	± 1 dot glide slope displacement
	Unusual throttle position for a significant period of time

Table 5-1. Wind shear recovery decision guidelines.[13]

The criteria in Table 5-1 are guidelines only; a brief digression from any of them may not result in a significant hazard to flight safety. On the other hand, you may need to initiate an immediate recovery procedure *before* any of them are exceeded in order to avoid a catastrophe. As pilot in command (PIC) you are responsible for the decision to abort a takeoff or conduct a go-around, but remember, it never hurts to err on the side of caution.

If you're unable to avoid an LLWS encounter after liftoff or on the approach, it's important that you act decisively, react promptly, and execute the recommended recovery procedure correctly—you may have as little as five seconds to recognize and respond to a wind shear encounter. Follow the guidance published in the aircraft manufacturer's flight or operating manual or your company's SOPs. If no specific guidance is available for your aircraft or operation, the *Pilot Windshear Guide* contains recommended procedures written for pilots of specific models of large commercial jet airplanes who encounter low-level microburst wind shear. Though type-specific, the principles are worth understanding by pilots of all aircraft—especially large transport turbojet airplanes.

Upon encountering LLWS in a transport category jet airplane as specified in AC 00-54—especially below 1,000 feet AGL—it is recommended that you immediately apply the maximum available thrust, pitch to the recommended climb attitude, and wait before changing aircraft configuration until terrain clearance is assured. You should use *full* thrust if terrain clearance is in doubt, and reduce it only if necessary to keep from overboosting the engine and once safety is assured. You should also wait to make configuration changes because climb performance can be significantly reduced by flap retraction, and to a lesser extent by the opening of gear doors during landing gear retraction. Finally, you should turn off the autopilot/FD and hand-fly the airplane if it is not designed to provide wind shear recovery guidance or if you cannot use it in pitch mode. Extensive flight testing has determined that a 15-degree nose-up pitch attitude is the optimum initial climb attitude for most of the models specified in FAA AC 00-54. If stick shaker (stall warning) occurs below that angle, it recommends lowering the nose just enough to stop it; if stick shaker doesn't activate, and altitude loss or vertical flight path angle is still unacceptable, then smoothly increase pitch attitude until stick shaker *intermittently* activates, then ease off until it stops. The goal is to obtain the maximum climb performance to keep the airplane flying long enough to successfully exit the wind shear and, of course, not stall or hit the ground. It is a delicate balance between pitching up to avoid ground contact and risking a stall, and pitching down to avoid a stall and risking ground contact. In fact, the *Guide* reports that many major wind shear accidents could have been averted if the crew members involved were better able to strike that balance. Several accidents occurred because pilots—including those aboard Delta Flight 191—erred on the side of avoiding stick shaker, and ended up contacting the surface.

The recovery guidelines in the above paragraph apply to specific Boeing, McDonnell Douglas and Lockheed transport category aircraft referenced in FAA AC 00-54; they do not necessarily apply to other aircraft, and some certainly don't apply to light aircraft. For example, a 15-degree pitch angle may be an inappropriate recovery attitude for the airplane you fly. Also, if you encounter LLWS in a light aircraft, do not increase pitch attitude until you get a stall warning; transport category aircraft have considerable climb capability at stall warning (stick shaker) while light aircraft do not. However, some of the *Pilot Windshear Guide*'s principles can still be useful even if you fly in a light aircraft. For example, the basic procedure is still the same: select full power, pitch to an optimal climb performance attitude, and wait until obstacle clearance is assured before changing aircraft configuration (but raise flaps from approach setting to a go-around or takeoff flap setting—consult the AFM). If no guidance is published in your aircraft's flight manual and you are uncertain as to what might be an optimal climb attitude, you can pitch to the normal go-around climb attitude, which, for most light airplanes, usually corresponds to the attitude that produces the best rate-of-climb speed (V_Y). Keep in mind, however, that wind shear may produce airspeed fluctuations, so it is important not to chase the airspeed.

Only when you are convinced that you have exited the wind shear should you reduce pitch attitude and accelerate to a normal cruise climb speed. Make sure to climb to a safe altitude as you determine your options to either hold and wait it out, attempt another approach (preferably to another runway), or divert to a more suitable airport. Just make sure you don't expose yourself to another LLWS encounter. The crew of Delta Flight 191 executed a successful recovery from their initial encounter with wind shear, only to be caught in it again as they continued the approach. And remember, always give a PIREP if you experience LLWS, making sure to indicate its location and the amount of airspeed loss or gain—your fellow pilots will appreciate it.

Though it is prudent to take precautionary measures and be able to execute a wind shear escape procedure if you encounter LLWS, as we have emphasized in this chapter, the best way to manage it is to avoid it. This involves obtaining a thorough weather briefing before flight and updates en route, and knowing how to recognize the signs of likely wind shear activity during flight. As is often the case with other hazardous weather scenarios, sometimes it simply involves waiting. It may be an old saw, but it is still true: *A superior pilot is one who uses his or her superior judgment to avoid those situations that would require the use of his or her superior skills.*

Helpful Resources

Pilot Windshear Guide (FAA AC 00-54) is an advisory circular published by the FAA that provides guidance primarily to flight crews of commercial transport category aircraft, but it has also been made available to a wider aviation audience. It is just one part of the two-volume *Windshear Training Aid*, developed by key members of the aviation industry in 1988. (www.FAA.gov/regulations_policies/advisory_circulars/index.cfm/go/document.information/documentID/22291)

The two-volume *Windshear Training Aid*, intended for commercial air carriers and developed by members of the airline and aerospace industry, consists of training documents, slides and video designed to enhance a pilot's understanding of wind shear. It is available in electronic, digital or print format (for a fee) from the National Technical Information Service (NTIS). Order numbers for *Volume 1: Overview, Pilot Guide, Training Program* and *Volume 2: Substantiating Data* are PB88-127196 and PB88-127204, respectively. (www.NTIS.gov)

Airbus offers an online safety library that provides several free flight operations briefing notes applicable to all aspects of safe commercial airline flight operations. One that is germane to this chapter is *Adverse Weather Operations: Windshear Awareness*. (www.AIRBUS.com/company/aircraft-manufacture/quality-and-safety-first/safety-library)

The International Civil Aviation Organization publishes the comprehensive *Manual on Low-level Wind Shear and Turbulence* (Doc 9817, AN/449).

We have already highlighted some of the risks involved in flying in the mountains and will do so again in future chapters. An aviation safety website developed by the Canadian Civil Air Search and Rescue Association (CASARA) contains a variety of excellent resources, including articles, videos and interactive programming to help pilots fly safe. (www.SmartPilot.ca) Many of its safety videos, including a set of short introductory videos on mountain flying (starring one of this book's authors), can also be seen on YouTube (www.YouTube.com):

- *Mountain Flying: Introduction*
- *Mountain Flying: Navigation*
- *Mountain Flying: Density Altitude*
- *Mountain Flying: Weather*
- *Mountain Flying: The 180 Turn*
- *Mountain Flying: Tips*
- *Mountain Flying: Lessons Learned*

Notes

1. Federal Aviation Administration, *Pilot Windshear Guide*, AC 00-54 (Washington, DC: FAA, November 1988).

2. Federal Aviation Administration, *A Windshear Avoided* (video) (Washington, DC: FAA, November 1995).

3. FAA, *Pilot Windshear Guide*.

4. International Civil Aviation Organization, *Manual on Low-level Wind Shear and Turbulence*, Doc 9817, AN/449 (Montreal: ICAO, 2005).

5. Flight Safety Foundation, "Killers in Aviation: FSF Task Force Presents Facts About Approach-and-Landing and Controlled-Flight-Into-Terrain Accidents," *Flight Safety Digest* (double-issue: November/December 1998; January/February 1999): 1–256. Available online at FlightSafety.org/archives-and-resources/publications/flight-safety-digest/flight-safety-digest-1998.

6. Boeing Commercial Airplanes, *Statistical Summary of Commercial Jet Airplane Accidents: Worldwide Operations, 1959–2009* (Seattle, WA: Boeing, July 2010).

7. United States National Research Council, *Low-Altitude Wind Shear and Its Hazard to Aviation: Report of the Committee on Low-Altitude Wind Shear and Its Hazard to Aviation, A Joint Study* (Washington, DC: U.S. NRC, 1983): 30.

8. FAA, *Pilot Windshear Guide*.

9. Federal Aviation Administration, *Computer Testing Supplement for Airline Transport Pilot and Aircraft Dispatcher,* Figure 144, FAA-CT-8080-7C. (Oklahoma City: FAA, 2005).

10. Federal Aviation Administration, *Aviation Weather Services*, AC 00-45G, Change 1 (Washington, DC: FAA, July 2010): 7–27. Available online at rgl.FAA.gov/Regulatory_and_Guidance_Library/rgAdvisoryCircular.nsf/0/D6A522C25E53CBF58625776F0050495C?OpenDocument.

11. Federal Aviation Administration, National Aviation Safety Data Analysis Center, *Review of Aviation Accidents Involving Weather Turbulence in the United States: 1992–2001*, Reference No. 04-551 (Washington, DC: FAA, August 2004).

12. D.A. Rhoda and M.L. Pawlak, *An Assessment of Thunderstorm Penetrations and Deviations by Commercial Aircraft in the Terminal Area*, NASA/A-2 (Lexington, MA: MIT Lincoln Laboratory, June 3, 1999).

13. FAA, *Pilot Windshear Guide*.

How High
Can You Fly?

6

HIGH-ALTITUDE FLIGHT

It was a warm sunny morning in April 1875 when Henri Theodore Sivel, Joseph Croce-Spinelli and Gaston Tissandier set out to break the altitude record in their gas-filled balloon. However, two hours into the flight at approximately 26,000 feet, all three French aeronauts aboard the *Zenith* lost consciousness. Tissandier awoke 40 minutes later to find their balloon rapidly descending. To correct for this, he cut away a bag of ballast; however, the balloon climbed and Tissandier again lost consciousness. An hour or so later he woke up to discover both his colleagues were still unconscious. They never awoke again—only Tissandier survived this 4½-hour flight that made it to an altitude of approximately 28,000 feet.[1]

It was a VFR evening in January 2003 when a pilot and three passengers embarked on a cross-country flight from Longmont, Colorado, to Las Vegas, Nevada. Unfortunately, they never made it to their destination. The engine of their Piper PA-28R quit due to fuel starvation, and the subsequent forced landing proved unsuccessful; the Cherokee Arrow crashed, killing all four aboard. According to the NTSB, the 140-hour private pilot seemed to have difficulty flying and navigating during the trip. The ARTCC controller gave the pilot numerous heading corrections—some by as much as 70 degrees—and at one point the pilot told ATC she was over the town of Montrose, but she was actually over a town located 35 miles away. It appears the engine failed because of improper fuel management (10 to 15 gallons of fuel likely remained in the unselected fuel tank at the time of the accident). Radar data indicates the pilot flew at an altitude of approximately 16,000 feet for an estimated 45 minutes, above 14,000 feet for one hour and 49 minutes, and above 12,500 feet for two hours and 17 minutes (NTSB Identification No: DEN03FA038).

What do both of these tragic events have in common? The pilots of both aircraft suffered from debilitating **hypoxia**—a lack of sufficient oxygen—as a result of flying at too high an altitude. Unless in a pressurized cabin or breathing supplemental oxygen, flying at high altitudes causes physiological and psychological impairment that inevitably leads to incapacitation. In the case of aviation's first known fatalities resulting from this phenomenon, the French aeronauts knew they needed oxygen to supplement their breathing; however, they didn't bring enough aboard to last the duration of the flight, and they failed to properly use what little they did have because of mental incapacitation brought on by hypoxia. Almost 130 years later it appears the Piper Arrow pilot wasn't aware of, or didn't fully appreciate, the dangers of hypoxia. The aircraft wasn't pressurized, it wasn't equipped with a supplemental oxygen system and there is no evidence the pilot took a portable supplemental oxygen unit on the airplane with her. Also, she may not have been aware of, or simply disregarded, the regulations that were implemented in the United States long after the deaths of Croce-Spinelli and Sivel: pilots are required to breathe **supplemental oxygen** at all times when flying above cabin altitudes of 14,000 feet MSL (13,000 feet in Canada). The pilot in this accident flew for almost two hours above that altitude.

Hypoxia adversely affects pilot performance in a number of ways. A quick search of NASA's ASRS database for the period covering the first decade of this millennium indicates how. Besides turning to wrong headings, leveling off at wrong altitudes, entering the wrong airspace, and dialing in the wrong frequencies on the radio, pilots suffering from hypoxia have started instrument approaches without an ATC clearance, unintentionally flown from VMC into IMC, flown 6,000 feet below the minimum IFR en route altitude (MEA) in mountainous terrain while in IMC, mismanaged fuel controls leading to engine failures, and misread altimeters by as much as 10,000 feet. Pilots have also reported losing their hearing (a pilot was unable to hear the cabin altitude warning horn), their memory (one pilot forgot an entire portion of the flight), and even their ability to move an arm to press the push-to-talk switch on the control column!

All of these examples fortunately had happy endings: the pilots involved lived to talk about them. Unfortunately, some others haven't. A search of the NTSB's accident database over the same period reveals that hypoxia accidents have a high lethality index, with 14 out of the 16 hypoxia accidents resulting in fatalities. Rather than making mistakes that only affected or could have affected the safety of flight (i.e., an incident), hypoxic pilots involved in accidents were either mentally unable to make the decisions needed to safely fly their aircraft or were rendered completely unconscious. That was the case for a private pilot flying from Glendive, Montana, to St. Paul, Minnesota. He was flying his Beech Baron at altitudes up to FL270 and was using oxygen. However, at some point in the flight he ran out of oxygen and became unconscious. Though military jets were dispatched to intercept and provide assistance, there was nothing they could do to rouse his attention. The airplane continued on autopilot, overflew St. Paul, and eventually crashed near Winfield, West Virginia. The NTSB determined the probable cause of the accident was the pilot's failure to ensure that an adequate supply of oxygen was available for the flight (NTSB Identification No: NYC06FA079).

The majority of hypoxic accidents involve private pilots and GA aircraft, but some involve experienced commercial pilots and scheduled airline operations. After an experienced ATP-certificated pilot transported them to jump altitude, all skydivers successfully exited the Beechcraft King Air B90; however, on the descent the airplane crashed into the Pacific Ocean, killing the pilot. This was the twelfth sport parachute jump of the day, and according to the skydivers, the previous two flights had been conducted at altitudes of at least 18,000 feet and this last jump was made at 20,000 feet MSL. Witnesses didn't observe the pilot using any supplemental oxygen and the NTSB couldn't find any evidence supplemental oxygen equipment was aboard. Investigators determined the probable cause of the accident was the pilot's incapacitation due to the effects of hypoxia from repeated flights to altitudes above 18,000 feet MSL without using supplemental oxygen (NTSB Identification No: LAX99LA190).

In a highly publicized occurrence, four passengers were on a chartered Part 135 flight from Orlando, Florida, to Dallas, Texas, when ATC lost communication with the crew of the Learjet 35. The airplane veered off course, and after several unsuccessful attempts to re-establish radio contact, ATC scrambled military jets to have a look. What they found was disturbing: no movement whatsoever of people inside the airplane, including the crew. The airplane continued en route (reaching a maximum altitude of 48,900 feet) under control of the autopilot until it ran out of fuel, entered a spiral dive and crashed near Aberdeen, South Dakota. The accident killed the captain, copilot and all four passengers, including U.S. professional golfer William Payne Stewart. The NTSB determined the crew became incapacitated as a result of its failure to receive supplemental oxygen after a loss of cabin pressurization. Investigators were unable to determine why the pressurization system failed (NTSB Identification No: DCA00MA005).

For the most part, hypoxia is prevented in most high-flying jets and commercial airliners by **cabin pressurization**. Hypoxia symptoms can be experienced at altitudes from 10,000 feet MSL and higher, so aircraft cabins are pressurized to a value lower than that. Though relatively rare, loss-of-cabin-pressure events, through system or structural failure, do occur. If a depressurization does occur, pilots of high-flying aircraft are trained to follow procedures that usually result in safe outcomes (e.g., use oxygen and conduct an emergency descent to a safe altitude). However, if crew members fail to recognize the decompression or are slow in taking the required corrective actions, they could succumb to hypoxia and jeopardize the safety of the flight. In the case of the Learjet 35 accident, it appears likely that the crew didn't recognize the depressurization in time to take the appropriate countermeasures.

Greece's worst aviation accident to date involved the failure to recognize a decompression. During the climb from Larnaca, Cyprus, on a scheduled flight to Prague in the Czech Republic, a warning horn activated in the cockpit of the Helios Airways Boeing 737. The captain mistook this for a takeoff configuration warning and didn't recognize it as the cabin altitude warning horn that activates above a cabin altitude of 10,000 feet. As they were troubleshooting the problem, and not knowing its true nature, the crew members fell victim to hypoxia. The airplane continued en route, flown by the autopilot and flight management system (FMS), until it ran out of fuel. All 121 onboard died as the airplane crashed into terrain near Athens.[2]

Hypoxia—An Insidious Threat

Oxygen, which is essential for life, is carried to the cells of the body by the hemoglobin in the red blood cells. Although the percentage of oxygen in the atmosphere remains relatively stable at approximately 21 percent, the pressure of this vital gas decreases as altitude is increased. For example, the pressure of oxygen exerted in the atmosphere at 18,000 feet MSL is approximately half (~3.15 inches of mercury [in Hg]) of that exerted at sea level (~6.3 in Hg). As altitude increases, the saturation of oxygen in the blood decreases until it reaches approximately 90 percent at 10,000 feet MSL. The physiological efficient zone lies below this altitude threshold, while the **physiological deficient zone** lies above it. Above 10,000 feet, hemoglobin's capacity to attract and hold onto oxygen is drastically reduced and a person begins to experience hypoxia.

Hypoxia is a deficiency of oxygen in the body's tissues (*hyp*=low; *oxia*=oxygen) and, as previously mentioned, it results in physiological and mental impairment. Aviation physiologists divide hypoxia into four general stages in which symptoms progress from mild to severe.[3] With the exception of night vision impairment above 5,000 feet (discussed in Chapter 7), there are usually no major effects during the indifferent stage, which generally occurs below 10,000 feet MSL. As you ascend within the first few thousand feet above 10,000 feet, or remain at or near 10,000 feet for any appreciable length of time, your body enters the compensatory stage of hypoxia in which it tries to compensate for the reduced partial pressure of oxygen by automatically increasing your heart and breathing rate. There is a tendency to hyperventilate, which is somewhat effective in combating the symptoms of hypoxia. (The symptoms of hyperventilation should not be confused with hypoxia itself.)

However, the negative effects of hypoxia are minimal only if exposure to these altitudes is of short duration. This is the reason regulations limit exposure to these altitudes for only 30 minutes (between 12,500 and 14,000 feet in the United States and between 10,000 and 13,000 feet in Canada). If enough time is spent at these altitudes, or if you ascend higher (up to approximately 20,000 feet MSL), the negative symptoms of hypoxia will be experienced during the disturbance stage. Total loss of consciousness, and even death, can occur during the critical stage if you remain at the upper end of these altitudes for a sufficient length of time or if you climb above 20,000 feet.

The type and severity of symptoms any one particular pilot may experience depend on a variety of factors including the altitude attained, the rate of ascent, duration at that altitude, and the frequency of exposure to these altitudes. It also depends on personal factors such as physical fitness, illness, medications, anemia and smoking. For many, the symptoms of hypoxia are not unlike those experienced by someone who has had several alcoholic beverages. Although the first symptom of hypoxia possibly could be unconsciousness, a pilot is more likely to experience one or several of the following symptoms: dizziness, tingling, headache, breathlessness, nausea, fatigue, lethargy, sleepiness, hot or cold sensations, tunnel vision or loss of visual acuity, reduced hearing ability, and poor coordination. Due to reduced oxygen in the red blood cells, cyanosis can also develop in the skin (e.g., turning purple in lips or fingertips, the blueberry effect).

The main organ affected by hypoxia, however, is the brain: Though it accounts for only about 2 percent of total body weight, it requires almost 25 percent of the body's oxygen intake to function effectively. Lack of sufficient oxygen to the brain can manifest itself in loss of alertness, difficulty in making mental calculations, memory loss (amnesia), moroseness, apathy, indifference, impaired judgment, overconfidence and belligerence.

Aviation physiologists distinguish between *good* and *bad* symptoms of hypoxia: A good symptom of hypoxia is one that is obvious and will alert you to its presence, such as a headache, nausea, dizziness, or tingling in the hands and fingers. A bad symptom is one that won't be so recognizable, such as euphoria. In fact, by far the most dangerous symptom of hypoxia is this sense of well-being. It is the same symptom that overcame the one remaining survivor of hypoxia's first victims on the *Zenith* over a century ago and continues to do so even today. Tissandier later recalled how he felt just before he first lost consciousness:

> About the height of 25,000 feet the condition of stupefaction which ensues is extraordinary. The mind and body weaken by degrees, and imperceptibly, without consciousness of it. No suffering is then experienced; on the contrary, an inner joy is felt like an irradiation from the surrounding flood of light. One becomes indifferent. One thinks no more of the perilous position or of danger. One ascends, and is happy to ascend.[4]

"One ascends, and is happy to ascend" truly describes the sense of well-being that will lull you into such a sense of security that you feel absolutely no need to do anything to counteract the threat. The experience is much like that of alcoholic intoxication: the more you drink—or the higher you fly—the greater the impairment.

One of the most dangerous problems with hypoxia is your inability to recognize its onset. It sneaks up on you, and if you are experiencing euphoria, it is even more insidious. At higher altitudes you have only a limited time period in which you are capable of making sound decisions and performing normal flight duties. **Effective performance time** (EPT), formerly referred to as the time of useful consciousness (TUC), decreases with altitude until it is less than a minute when flying at altitudes above 30,000 feet. This means you have less than a minute to take proper measures to stop hypoxia's effects. According to the FAA's advisory circular on high-altitude flight (AC 61-107), EPTs (or "TUC times") range from about 20 minutes at 18,000 feet to 20 seconds at 40,000 feet.[5] EPTs are influenced by diet, exercise, illness and fatigue, and are shorter for smokers and those who are not physically fit.

Cabin Pressurization

Hypoxia is prevented in non-pressurized aircraft through use of onboard fixed or portable oxygen systems. As mentioned previously, in most high-flying aircraft, hypoxia is prevented through cabin pressurization. Most pressurization systems compress the air in the cabin by directing compressed intake air from the engines to the cabin, and control the air's expulsion from the cabin through the adjustment of outlet valve settings. A cabin altimeter (which indicates cabin pressure expressed as an equivalent altitude) and cabin rate-of-climb/descent indicator is used by pilots to monitor the cabin altitude and rate of altitude change within the cabin. 14 CFR §25.841 requires that the air in the cabin be compressed to a cabin pressure altitude of not more than 8,000 feet when the airplane is flying at its maximum operating altitude. This is a compromise between maintaining safety and comfort (i.e., avoiding hypoxic cabin altitudes above 10,000 feet) and keeping increased weight and subsequent reduction of aircraft performance to a minimum.

Sudden Decompressions

As previously mentioned, although cabin depressurizations are relatively rare events, they do occur. Citing data provided by Airbus and Boeing, the FAA reports in its final rule regarding extended operations (ETOPS) of multi-engine airplanes that between 1959 and 2001, there were almost 3,000 depressurization events in transport category airplanes weighing more than 60,000 pounds. For Boeing airplanes alone, there were 73 events between 1980 and 2000.[6] The FAA estimates the odds of experiencing a loss-of-cabin-pressure event in a large transport category airplane to be between one in a million and one in 10 million per flight hour. The vast majority of these depressurization events were the result of pressurization system failures (pressure controllers, valves, etc.), followed by structural failures (door seals, etc.) and very rarely by **uncontained engine failures**. Using Airbus data, the FAA estimates the probability of experiencing a decompression due to pressurization system failure alone is about 3.5 decompression events for every million flights.[7]

Some decompressions are gradual while others are more sudden. In a gradual decompression, it usually takes longer than 10 seconds for the cabin pressure to equalize with that of the outside environment. **Sudden decompressions** are classified as either rapid or explosive: A rapid decompression takes 0.5 to 10 seconds to occur, while an explosive one takes less than 0.5 seconds. With the latter, air in the lungs is unable to escape as fast as the air in the cabin and lung damage could occur, especially if a person's airway is closed due to swallowing or holding their breath. According to incident data, the majority of decompressions are gradual, few are rapid, and fewer still are explosive.[8]

Airplanes with larger cabins are less likely to experience explosive decompressions, because given the same circumstances, airplanes with larger cabin volumes take longer to decompress than those with smaller volumes. According to FAA Advisory Circular 61-107 on high-altitude flight, the cabin volume of a Boeing 747 is about 59,000 cubic feet compared to a Learjet at about 265 cubic feet, yielding a cabin volume ratio between the two of 223 to 1. Therefore, given the same type of depressurization, the Learjet will decompress 223 times faster than the B-747.

Depending on the rate of pressure equalization, the phenomena experienced during the decompression itself could range from relatively negligible to downright frightening. During a sudden—rapid or explosive—decompression, you will hear a loud bang or noise, feel the rushing wind, and see flying dust and debris as it escapes from the cabin. You also will likely experience the uncomfortable sensations of air escaping from the various orifices of your body, and any excess pressure in your ears will quickly clear. Fog could form from water vapor condensing as a result of the rapid expansion of air. As warm cabin air is displaced by cold outside air, the cabin air temperature will also drop considerably, causing significant discomfort. Finally, passengers will feel anxious and fearful, and so will you as you quickly try to ascertain what just happened. It should be noted that very rarely has someone involuntarily exited the airplane because they were located near the opening and were unsecured by a safety harness.

As harrowing as it might be perceived by crew or passengers, in the vast majority of decompressions the major threat to health and safety does not occur during the decompression event itself, but after it takes place—and that threat is hypoxia. Since air escapes from the body's tissues very quickly during a sudden decompression, EPTs are reduced by about half. That means you (and your fellow crew members), who would normally have about one to two minutes to take the required measures to secure your safety, have only about 30 to 60 seconds following a sudden decompression at 30,000 feet.

Types of Hypoxia

Up to this point we have been describing the type of hypoxia aviators are most likely to encounter—inadequate oxygen due to flying at high altitudes, also known as **hypoxic hypoxia**. There are three others that you should be aware of: **hypemic**, **stagnant** and **histotoxic hypoxia**.

Hypemic (or anemic) hypoxia is caused by insufficient oxygen-carrying capacity in the blood due to low blood iron, loss of blood (including loss from blood donations) or carbon monoxide (CO) inhalation. Oxygen is transported in the blood by combining with hemoglobin to form oxyhemoglobin. However, since CO—inhaled from cigarette smoke or exhaust fumes—has an affinity for hemoglobin that is 200 to 300 times greater than that of oxygen, it competes with oxygen, effectively displacing it, combining with hemoglobin to form carboxyhemoglobin. This reduces the blood's ability to carry oxygen to the tissues. Cigarette (or cigar) smokers, therefore, are more susceptible to hypoxia. Experts say that people who smoke regularly are at a physiological altitude of about 5,000 feet when they are at sea level.

Stagnant hypoxia occurs when blood flow is reduced due to circulatory system problems (i.e., stagnation of blood). Heart failure, blood vessel restrictions and excessive G-force can cause this type of hypoxia.

Histotoxic hypoxia develops when the body's cells fail to efficiently use the oxygen delivered to them. The supply of oxygen from the blood is adequate, but the tissues—due to alcohol, drugs or poisons—are unable to use the oxygen: the tissues (*histo*) themselves become toxic.[9]

Avoiding Hypoxia

It should be obvious that humans were not meant to fly at high altitudes without proper protection. Flying in a pressurized cabin or using supplemental oxygen allows you to fly higher, but you still need to know what to do if these systems fail. Therefore, whether or not your aircraft is equipped with these systems, it is essential you know what measures to take to prevent hypoxia from sneaking up on you.

Avoid Hypoxic Altitudes

If you have no need to fly above 10,000 feet MSL (or 5,000 feet at night—see discussion of hypoxia's effects on vision in Chapter 7), then don't. This will keep you from experiencing the negative symptoms of reduced oxygen. Keep in mind that hypoxia is not just a function of altitude attained but also duration at that altitude: Prolonged flight at or near hypoxic altitude thresholds can cause hypoxia.

On one occasion Dale Wilson, one of this book's authors, conducted a 1,000-mile cross-country flight in a small airplane. After almost four hours at only 9,500 feet MSL, he experienced hypoxia's effects. He remembers looking at his passenger, who was also a pilot, with a wide grin and saying, "Hey, I'm feeling pretty good right now. I think I'm getting hypoxia." His passenger looked back at him with an equally large grin and said, "Yeah, me, too." "Do you think we should descend?" Dale said, still feeling pretty happy with that big smile on his face, to which his pilot-friend replied, "Yeah, probably a good idea." It was insidious—it snuck up on them—but fortunately, they were able to detect it.

Be Careful at Marginal Altitudes

Oxygen use in the United States is not required for pilots flying under Part 91 regulations when flying at or below 12,500 feet MSL. However, it is required after flying for more than 30 minutes between 12,500 and 14,000 feet MSL and at all times when flying above 14,000 feet MSL. In Canada, it is not required when flying at or below 10,000 feet MSL, but is after flying more than 30 minutes between 10,000 and 13,000 feet MSL and at all times when flying above 13,000 feet MSL. However, pilots may erroneously think that because it's legal to operate an aircraft up to these altitudes without using supplemental oxygen, it must be safe. Remember, at these altitudes the body tries to compensate for reduced oxygen by increasing both the heart and breathing rate, and it can only do so for a limited period of time before the symptoms of hypoxia take hold. To avoid its insidious effects, you should limit time at these marginal altitudes. If you have to remain there because of terrain or weather, be on the lookout for hypoxia's signs, and if present, descend to a lower altitude or begin to supplement your breathing with oxygen.

One more point: Pilots tend to hyperventilate (*hyper*=over; *ventilate*=breathe) at these altitudes as the body attempts to compensate for reduced oxygen. As mentioned, the symptoms can be confused with those of hypoxia. If you are experiencing symptoms and are unsure of their origin, deliberately slow your rate of breathing, and if hyperventilation is the culprit, then the symptoms should clear up within a few minutes. If you have oxygen available, begin using it, and if hypoxia is the cause, the symptoms should clear up within a few breaths.

Consider Hypoxia Training

Not everyone experiences hypoxia the same way. Some encounter easily recognizable symptoms such as headaches, nausea, or tingling of the skin; others experience not-so-noticeable symptoms such as indifference, apathy or euphoria. Participating in altitude chamber or other reduced-oxygen training gives you the ability to identify your own unique symptoms in a controlled environment before you are exposed to hypoxia in the real world of flight. That way, if hypoxia sneaks up on you, you will be ready since you know what to look for ahead of time. The FAA provides physiology training to civilian pilots, including simulation to FL250 in a hypobaric chamber, at the FAA's Civil Aerospace Medical Institute (CAMI) in Oklahoma City. Up until recently, it also offered such training at selected military installations throughout the country. There are universities and private contractors who provide this training to pilots as well. Incidents have occurred in which crew members who've participated in such training have been incapacitated from hypoxia, so participating in altitude chamber training to discover your own unique symptoms won't automatically prevent you from succumbing to hypoxia—but it should reduce your odds.

Decrease Your Susceptibility

Certain factors increase one's susceptibility to hypoxia. Some of these include smoking, use of alcohol, and use of over-the-counter (OTC) medications such as antihistamines and pain relievers. Be aware that the underlying conditions requiring such medications can also reduce your tolerance. Of course, anemia and loss of blood through **blood donations** significantly reduce the blood's ability to carry needed oxygen to the body's tissues. The FAA's *Pilot's Handbook of Aeronautical Knowledge* states that it takes several weeks for blood levels to return to normal after donating blood; therefore, during that time your body will not be able to function at its peak capacity.[10] This may not be a problem at sea level, but the negative effects of reduced oxygen in the blood at altitude are likely to be significant. Nothing in the regulations prohibits you from donating blood, but the TC *AIM* recommends that active pilots should not donate blood at all. If you do donate, it recommends that you wait at least 48 hours before resuming flying activities.

Follow the Regulations

We previously summarized the basic altitude restrictions for GA pilots flying without oxygen. Simply following these regulations can go a long way to keeping you out of trouble. It sure could have for the 140-hour private pilot and her three passengers (mentioned earlier) trying to get to Las Vegas, as well as many others who've been killed by hypoxia. A search of the NTSB's accident database for a recent 10-year period shows that pilots in about half of these types of accidents failed to comply with these basic rules.

Of course, the regulations are even more stringent for commercial passenger-carrying flights. Thresholds for mandatory oxygen use are lowered to *cabin altitudes* of 10,000 feet for flight crews flying under Part 121 (scheduled airline) and to no longer than 30 minutes between cabin altitudes of 10,000 and 12,000 feet for pilots flying under Part 135 regulations (commuter and on-demand). These rules also apply should an aircraft lose pressurization. There is an extensive list of other regulations regarding oxygen requirements for flight crew, cabin crew and passengers; you must fully understand the applicable federal aviation regulations (FARs), or Canadian aviation regulations (CARs) in Canada, if you fly commercially.

Know Your Oxygen System

Whether you need oxygen to take advantage of the benefits of high-altitude flight (e.g., higher true airspeed, faster tailwind, etc.) in an unpressurized aircraft, or need it in the event of a rapid decompression in a pressurized cabin, it's crucial that you receive training and fully understand the **oxygen system** in your aircraft. There are a variety of fixed and portable systems. The most common for light aircraft is a portable tank containing oxygen, valves, regulator, pressure and flow gauges, hoses and mask(s). There are also fixed installations consisting of fixed oxygen tanks that can be filled from an external valve with a built-in system of tubes to deliver the oxygen to the pilots' and passengers' masks. Whichever system is installed, you should be thoroughly familiar with it.

A commonly used acronym to check for proper functioning of the equipment is the **PRICE check**. You should first check the *pressure* gauge to confirm there is enough pressure and an adequate oxygen supply. Next is the pressure *regulator*, which reduces pressure and regulates oxygen flow. Confirm the system is set properly and delivering oxygen. You can check the oxygen flow rate on the *indicator*. Ensure that all *connections*, including face mask, are not leaking or twisted and are properly connected. Finally, you should have an *emergency* supply of oxygen in the cockpit (regulations for airline flights require this).

It's important to ensure the system is working properly, but it's just as important to determine if you have enough oxygen for the proposed flight. Keep in mind that the higher you go, the more oxygen you will use and the more frequently you will need to check your supply. Most systems increase the amount of oxygen delivered to the mask with altitude until at approximately 34,000 feet MSL, at which point they provide 100 percent oxygen to the user. It is not known whether the pilot of the turbocharged Beech Baron conducted these checks or not, but if he did, it might have prevented his incapacitation due to hypoxia. As mentioned earlier in this chapter, the aircraft overflew its destination of St. Paul, Minnesota.

Despite efforts to establish contact with the pilot, the airplane eventually descended and crashed in Winfield, West Virginia. The NTSB determined the cause of the crash was the "pilot's inadequate preflight preparation to ensure an adequate supply of supplemental oxygen, and his inadequate in-flight planning and decision making, which resulted in exhaustion of his oxygen supply" (NTSB Identification No: NYC06FA079).

When servicing oxygen equipment you should make sure you use **aviator's breathing oxygen**. It is purer than oxygen used for other purposes and is also drier, which precludes the possibility of the system freezing at the cold temperatures found at higher altitudes. A Cessna 337 entered an uncontrolled descent and crashed near Hickory, Pennsylvania, after the pilot lost consciousness at 27,700 feet. According to the surviving passenger, the portable oxygen system used by the Skymaster pilot appeared to be functioning properly. However, according to the NTSB investigation, the bottle was not filled with 100 percent oxygen but with compressed air that provided only 21 percent oxygen (NTSB Identification No: IAD97FA060).

Finally, be very careful with oxygen. Since it is a necessary component of combustion—along with heat and fuel in what is known as the fire triangle—it goes without saying that you shouldn't smoke in its presence. You should also avoid using flammable oil- or petroleum-based products (the "fuel" in the fire triangle), such as lip balms, when using oxygen.

Know Your Pressurization System

If you fly in a pressurized cabin it is vital to understand not only how the environmental control system (ECS) in your aircraft works but also how to properly operate it. Several self-induced decompressions have occurred because flight crews have improperly managed this system. That was the case in a rapid decompression to FL330 in an American Trans Air Boeing 727 near Indianapolis in 1996, when the flight engineer (FE) inadvertently opened the pressurization outlet valve when troubleshooting the system. The captain, FE and a flight attendant all initially lost consciousness due to hypoxia. But the catastrophe was averted by a sharp FO who, although he had only 10 hours of initial operating experience (IOE) in the airplane, immediately donned his oxygen mask and conducted an emergency descent to a safe altitude (NTSB Identification No: CHI96IA157). In the fatal Helios Airways accident mentioned previously, the crew failed to recognize that the cabin pressure mode selector was left in the manual (MAN) mode after an earlier ground check of the system by maintenance personnel. It should have been set to AUTO. In MAN mode the system does not automatically control the desired pressurization for flight but requires the crew to manually adjust the cabin pressurization settings.

Finally, in several cases crew members have confused the cabin altitude warning horn—designed to activate when cabin pressure exceeds 10,000 feet—for some other warning. In some airplanes, especially older ones, this is understandable since the same horn is used to signify other anomalies, most commonly a takeoff warning. This is what happened in the Helios Airways accident: the captain mistook the cabin altitude warning for a takeoff warning. For many airplanes this warning is not tied into the master warning/caution system, and the horn is the only warning of excessive cabin altitude; the crew must look at the cabin altitude instruments to confirm if the horn signifies an excessive cabin altitude. Numerous occurrences, including the Helios accident, prompted the FAA to issue a *SAFO* for Boeing 737 aircraft, advising crew members to be aware of this confusion and to immediately don their oxygen masks as soon as the warning horn sounds in flight while they troubleshoot the problem.[11]

Follow Proper Decompression Procedures

Finally, in the unlikely event of a sudden or even gradual cabin decompression, you need to know what to do. The crucial point to remember is you will become incapacitated if you don't act quickly. Several accidents and incidents have occurred because pilots did not follow the proper procedures in a timely manner. As soon as the decompression occurs, or is recognized, you should follow the recommended emergency procedures published in the approved AFM for your aircraft. The general procedure is for you and your crew to immediately don your oxygen masks and ensure they are delivering a 100 percent supply of oxygen, and then initiate a maximum rate emergency descent—spoilers, landing gear and flaps deployed as appropriate—to 10,000 feet MSL (or the minimum safe IFR altitude if higher). As a result of the Payne Stewart crash, the NTSB looked at 129 ASRS reports over a 10-year period and found many pilots did not immediately don their oxygen masks while troubleshooting cabin pressure altitude problems. In its *Safety Recommendation* (A-00-109 through -119), the NTSB stresses the absolute necessity of donning oxygen masks as soon as the cabin altitude warning horn sounds or when the decompression is first recognized.[12]

There are also other reasons to quickly initiate an emergency descent: It precludes those who, for whatever reason, might not be successful in receiving oxygen from their masks from experiencing serious hypoxia; it reduces the demand for oxygen since regulations require only a limited supply of oxygen be carried for passengers; and, if the decompression occurs above 25,000 feet, it reduces the chance of you and your passengers experiencing **decompression sickness** (DCS).

Analogous to carbon dioxide gas rising to the surface once the pressure is released after a can of soda pop is opened, DCS occurs when dissolved nitrogen comes out of solution in the body's tissues and forms bubbles as a result of a drop in barometric pressure, either from a sudden decompression or from flying at too high of an altitude. The generally accepted threshold for DCS is 25,000 feet, but symptoms can occur as low as 18,000 feet MSL. Debilitating physiological symptoms result from nitrogen gas bubbles forming in the blood, skin, lungs, nervous system and the brain. But the most common symptom—**the bends**—causes severe joint pain from nitrogen bubbling into the joints and muscles. This is why pilots and passengers should avoid flight within 24 hours after **scuba diving**: the deeper one dives, the greater the quantity of nitrogen that goes into solution. If there is insufficient time to allow the excess nitrogen to escape normally through the lungs, subsequent exposure to even low flight altitudes can lead to DCS. If you plan to fly below 8,000 feet MSL, both the FAA *AIM* and TC *AIM* recommend waiting at least 24 hours after a dive if it involved a controlled ascent (i.e., decompression-stop dive) and at least 12 hours if it did not require a controlled ascent. Due to the possibility of a sudden decompression, they also recommend waiting at least 24 hours if flying above 8,000 feet MSL, regardless of the type of dive.

More than a century ago, Paul Bert—a pioneer in the field of altitude physiology and inventor of the first hypobaric chamber—warned the crew of the *Zenith* that they needed more oxygen for their proposed flight. Despite this admonition, they decided to launch and the rest of that fateful day is history. Unfortunately, there are still pilots who either are unaware of the hazards of high-altitude flight or ignore the precautions necessary to safely manage them. If you've fully grasped the content of this chapter, you can be thankful that you're not one of them. To rephrase an old adage, "You've learned from the mistakes of other pilots— those who've gone before you in succumbing to hypoxia—because you know you won't live long if you make them yourself."

Helpful Resources

Chapter 16 (*Aeromedical Factors*) in the FAA *Pilot's Handbook of Aeronautical Knowledge* (FAA-H-8083-25) provides an excellent overview of hypoxia. (www.FAA.gov/regulations_policies/handbooks_manuals/aviation/pilot_handbook/)

Aeromedical Training for Flight Personnel. A great reference providing in-depth coverage of a variety of physiological aspects related to flying, including hypoxia and DCS; it is published by the U.S. Department of the Army and is available to the public through booksellers and from other publishers.

The FAA's Civil Aerospace Medical Institute (CAMI) provides the following free videos dealing with various physiological aspects of flight, including high-altitude flight. (www.FAA.gov/about/office_org/headquarters_offices/avs/offices/aam/cami/library/online_libraries/aerospace_medicine/aircrew/physiologyvideos/english/)

- *Altitude Induced Decompression Sickness*
- *Flying and Hypoxia*
- *Respiration and Circulation*
- *Understanding Aviation Oxygen Equipment*
- *The Ups and Downs of Cabin Pressurization*

Aircraft Operations at Altitudes Above 25,000 Feet Mean Sea Level or Mach Numbers Greater Than .75 (FAA AC 61-107). This advisory circular provides physiological and aerodynamic information for pilots upgrading to complex, high-performance aircraft capable of operating at high altitudes and high airspeeds. (www.FAA.gov/regulations_policies/advisory_circulars/index.cfm/go/document.list/parentTopicID/119)

Notes

1. John Mackenzie Bacon, *The Dominion of the Air: The Story of Aerial Navigation* (New York: Cassell & Company, 1902): 223–228.

2. Mark Lacagnina, "Missed Opportunities," *AeroSafety World* 2.1 (January 2007): 18–24. Available online at FlightSafety.org/aerosafety-world-magazine/past-issues.

3. U.S. Department of the Army, *Aeromedical Training for Flight Personnel*, Training Circular No. 3-04.93/FM 3-04.301 (Washington, DC: U.S. Dept. of the Army, August 31, 2009).

4. Bacon, *Dominion of the Air*, 226.

5. Federal Aviation Administration, *Aircraft Operations at Altitudes Above 25,000 Feet Mean Sea Level or Mach Numbers Greater Than .75*, AC 61-107B (Washington, DC: FAA, March 29, 2013).

6. Federal Aviation Administration, "Extended Operations (ETOPS) of Multi-Engine Airplanes; Final Rule," *Federal Register* 72.9 (January 16, 2007): 1807–1887.

7. Ibid.

8. Federal Aviation Administration, *Interim Policy on High Altitude Cabin Decompression*, Memo No. ANM-03-112-16 (Washington, DC: FAA, March 24, 2006).

9. U.S. Department of the Army, *Aeromedical Training*.

10. Federal Aviation Administration, *Pilot's Handbook of Aeronautical Knowledge*, FAA-H-8083-25A (Washington, DC: FAA, 2009).

11. Federal Aviation Administration, "Boeing 737 (B-737) Cabin Altitude Warning Horn Confusion," *Safety Alert for Operators*, SAFO 08016 (Washington, DC: FAA, July 7, 2008).

12. National Transportation Safety Board, *Safety Recommendation*, A-00-109 through -119 (Washington, DC: FAA, December 20, 2000).

Don't Be Caught in the Dark

7

FLYING AT NIGHT

It was a pleasant December evening in Miami. The temperature was warm, the sky was clear and the visibility was unrestricted. The crew members of Eastern Air Lines Flight 401 were preparing for landing at Miami International Airport when the nose landing gear green light, which would indicate that the nose gear was down and locked, failed to illuminate. The captain recycled the landing gear, but to no avail—the light still did not illuminate. The flight received vectors from ATC to circle northwest of the airport while the crew tried to ascertain if the problem was a genuine gear malfunction or just a burnt-out light bulb. The FO selected the autopilot to maintain 2,000 feet; however, while the crew was preoccupied with troubleshooting the problem, someone (probably the captain) inadvertently bumped the flight controls. Now in the control wheel steering (CWS) mode, the autopilot dutifully flew the Lockheed L-1011 TriStar in a gentle descent that, unfortunately, went unnoticed by the crew. It wasn't until seven seconds before impact that the FO first noticed something was amiss saying, "We did something to the altitude. We're still at two thousand, right?" Six radio altimeter beeps sounded in the cockpit to warn the crew, but it was too late—the jumbo jet descended into the swamp of the Florida Everglades, killing 101 of the 176 people aboard (Report No: NTSB-AAR-73-14).

Flying an aircraft at night is often a pleasant experience. The ride is usually smoother, fewer aircraft are in the sky, the radio is quieter—and the view outside the cockpit is sometimes simply spectacular. However, as seen in the Florida Everglades accident, it is not without risk. A qualified and experienced flight crew allowed an otherwise serviceable aircraft to continue flight toward terrain until it was too late to recover. It was not IMC or poor visibility that precluded the pilots from seeing the approaching terrain—good VMC prevailed. Rather, it was the darkness of night. All else being equal, it is likely that this accident, along with many others like it, would never have occurred had it been daylight.

As is true for driving an automobile, flying at night poses an elevated level of risk compared to flying during the day. According to the National Safety Council, your chance of being killed in a motor vehicle accident is three times greater at night than during the day.[1] According to the AOPA Air Safety Institute, the odds of experiencing a fatal accident in a GA aircraft are more than doubled when flying at night—and increase even more depending on the type of flight operation (e.g., night IMC has the highest lethality index, increasing the probability fivefold).[2] Its threat to safety is so noteworthy that the NTSB has issued a *Safety Alert* warning pilots about the risks of *visual* flight operations at night—whether conducted under VFR flight or during the visual portions of IFR flight.[3]

Hazards That Lurk in the Dark

Just as certain types of motor vehicle accidents occur more frequently at night, certain kinds of aircraft accidents also are more prone to occur after dark. These occur during all phases of flight, from taxi and takeoff to the approach and landing. A closer examination of these will help you better understand the nature of this hazard.

Ground Operations at Night

In Chapter 1, we pointed out that reduced optic flow in peripheral vision leads pilots to taxi too fast, making it harder to accurately navigate to and from the runway during the hours of darkness. The **sea-of-blue effect** created by the maze of blue taxiway lights compounds the problem, making it difficult to differentiate between taxi routes. This is especially true if you fly smaller GA aircraft with lower eye-to-wheel heights. This is one of the many contributing factors that increase the likelihood of colliding with other aircraft during ground operations. In fact, as explored in Chapter 1, most accidents resulting from runway incursions occur in conditions of reduced visibility either in fog, the darkness of night, or both. For example, of the six major fatal runway incursion accidents in the United States during the 1990s, five occurred after dark (or at dusk) in VFR weather conditions.[4] The flight crews of Eastern Air Lines' B-727 at Hartsfield in 1990 and USAir's B-737 at Los Angeles in 1991 reportedly didn't see the other airplanes stopped on the runway ahead of them until the nosewheels of their airplanes touched down after landing and the airplanes' landing lights illuminated the other aircraft. At that point it was too late, unfortunately, for either of them to avoid a fatal collision (Report No: NTSB/AAR-91/03; NTSB/AAR-91/08).

Takeoff and Climb in the Dark

Flying at night can pose serious hazards during the takeoff phase of flight. In Chapter 1 we referenced a NASA study that found a surprisingly high number of wrong-runway and taxiway takeoffs conducted by commercial airline pilots; it turns out that the darkness of night contributed to many of them.[5] A more recent search reveals the surprising regularity of these incidents during the first decade of this new millennium. For example, a B-737 reached the speed of 60 knots before the captain aborted takeoff at Phoenix Sky Harbor Airport after realizing he was attempting a takeoff on Taxiway Foxtrot (ASRS Report No: 622348). Several jetliners at other airports have executed RTOs after realizing the same thing. For example, a Cessna Citation nearly collided with a regional jet when it departed Taxiway Uniform at Richmond International Airport in Virginia, and the crew of a McDonnell Douglas MD-80 initially attempted a takeoff in Las Vegas using the runway edge lights as centerline lights (ASRS Report No: 606541; 513080). All these incidents occurred in VMC, but after sundown. Thankfully, none resulted in an accident. As discussed in Chapter 1, that was not the case for Comair Flight 5191, a Bombardier Challenger 600 carrying three crew members and 47 passengers that crashed during takeoff after the crew attempted to depart on Runway 26—a runway with insufficient takeoff distance—instead of Runway 22. The accident occurred about one hour before sunrise during night VMC with no illumination from the moon (Report No: NTSB/AAR-07/05).

A serious hazard also occurs after liftoff when the airplane is in the initial climb phase of flight, especially if that climb is made into black-hole conditions. These conditions occur in VMC and are particularly pronounced on dark nights (moonless and/or overcast sky) when the departure area off the end of the runway is devoid of ground lights and no horizon is discernible (as is often the case over water, for example). According to the TSB, more than three-quarters of night takeoff accidents in Canada occur in **dark-night conditions**.[6] If you rely solely on external visual cues after takeoff on dark nights, you may fail to establish the required climb gradient to clear unseen rising terrain ahead, or you may experience a false sensation of a nose-up attitude. This illusion could in turn lead you to pitch the nose down, causing the airplane to descend into the ground or water.

The physiological factors that contribute to this **false climb illusion** will be discussed in greater detail in Chapter 9, but for now it is enough to know that in the absence of adequate outside visual references at night, the vestibular apparatus in the inner ear is unable to distinguish between straight-line accelerations and head-up or head-down tilt. Therefore, accelerating down the runway on takeoff, or even during a shallow climb after liftoff, has the same physiological effect as tilting your head up. Designers of full-motion visual simulators actually take advantage of this physiological trait to simulate acceleration; on takeoff the front of the device is tilted upward, tricking pilots into thinking they are accelerating along a level runway. It should be apparent that the more rapid the acceleration (as would be the case with higher-powered aircraft), the more pronounced the illusion.

Although the hazard of the **dark-night takeoff accident** has been known for almost as long as pilots have been flying at night, it is still occurring. The false climb illusion was specifically cited as a causal factor in a dark-night takeoff accident involving a Part 135 cargo flight in 1996 in Kamuela, Hawaii, which claimed the life of a commercial pilot holding an ATP certificate (NTSB Identification No: LAX96FA103). More recently, an 18,000-hour commercial pilot died when his Beechcraft Baron G58 impacted the waters of Lake Erie after departing a lakeshore airport in Cleveland, Ohio, and a commercial pilot flying a Cessna 340 lost his life in a dark-night takeoff accident in Bishop, California, even though the visibility was 10 miles and the sky was clear (NTSB Identification No: CHI08FA066; LAX03FA254).

Cruising in the Dark

The inability to adequately see outside at night is also a hazard during the en route, or cruise, portion of flight. It's sometimes difficult to know your exact location or to detect outside threats such as high terrain or adverse weather when flying in the dark.

Geographic Disorientation

If you are flying at night under visual flight rules (NVFR) but are not using radio navigation to confirm your position, there is a greater chance of experiencing geographic disorientation (i.e., getting lost). If there is not enough light to visually confirm your position with reference to natural or human-made landmarks, it stands to reason that it will be more difficult to navigate at night, especially on a dark night.

Terrain Detection

It's also difficult to visually detect terrain in the dark, even in clear weather. Accidents where pilots fly into unseen mountainous terrain on clear dark nights still occur. One of the most publicized of these claimed the lives of eight members of country singer Reba McEntire's band. Flying below the San Diego terminal control area (now called Class B airspace) on a clear moonless night with a visibility of 10 miles, while waiting for an IFR clearance, the flight crew flew the band's Hawker Siddeley 125 under controlled flight into mountainous terrain near Brown Field Municipal Airport. Unfortunately, the flight and medical crews of an air ambulance Learjet suffered the same fate in almost the same location 13 years later. While waiting for an IFR clearance and maintaining VFR flight below a cloud deck, the aircraft crashed within 1.5 miles of the 1991 crash site, killing all onboard (NTSB Identification No: LAX91FA132; LAX05FA015).

Weather Detection

Just as it is difficult to see terrain in the dark, it is also difficult to see adverse weather. As noted in Chapter 4, VFR pilots are at least three times more likely to experience a VFR-flight-into-IMC accident at night. Not only are these types of accidents one of the leading causes of fatalities in aviation, the AOPA Air Safety Institute found that VFR-into-IMC accidents at night were the number one cause of fatal GA weather accidents in the United States.[7]

Volcanic Ash Detection

Though the probability of flying into **volcanic ash** is low, the consequences of such action can be catastrophic. The first recorded instance of a total powerplant failure resulting from flight into volcanic ash occurred during the hours of darkness in clear weather conditions. It happened to a British Airways B-747 while in cruise flight over Indonesia in 1982. The airplane lost 25,000 feet and glided for more than 15 minutes before the crew was able to restart three engines and safely land at a nearby airport in Jakarta. A month later, a Singapore Airlines B-747 flew into volcanic ash released from the same volcano, resulting in the loss of three of its four engines. Fortunately, the crew was able to restart one engine and land safely at a nearby airport after losing nearly 8,000 feet of altitude.[8] Both of these incidents occurred at night, a condition that makes it virtually impossible to detect volcanic ash.

Approach and Landing at Night

During the first decade of this century, more than a third of the world's fatal commercial jet airplane accidents occurred during the final approach and landing phases of flight, yet these phases only occupied about 4 percent of flight time.[9] Accident data confirm the darkness of night adds considerable risk. For example, one study found that 39 percent of fatal approach and landing accidents during a 15-year period in the United Kingdom occurred at night, leading investigators to conclude the nighttime accident rate to be three times greater than during the day.[10] Another study of U.S. Navy/Marine Corps accidents that occurred over a 16-year period found a significantly greater incidence of landing mishaps during the evening and at night.[11] And finally, another safety study found that one-half of commercial airline CFIT accidents occurred at night, and 70 percent of those occurred during the approach-to-landing phase of flight.[12]

Airport and Runway Detection

Before landing at an airport you must first be able to find it. This is not always easy at night, especially if you are relying solely on external visual cues for guidance. In one year alone, 20 U.S. commercial air carrier flight crews conducted approaches to, or landed at, the **wrong airport**. Over a six-year period, 21 percent of these types of incidents occurred at night, mostly in clear visual conditions.[13] In one instance, four crew members and 26 passengers died when their Convair CV-640 crashed 28 miles short of its destination while on approach to a site mistaken for the airport; it was a clear moonless night.[14] In the early 1990s, a DC-8 crashed nine miles short of Mallam Aminu Kano International Airport in Nigeria after the flight crew mistook army barrack lights for those of the runway. The same unfortunate mishap occurred nine months later when a B-707 crashed within 100 yards of the DC-8 accident site![15] More recently, the crew of a B-727 cargo flight mistakenly landed on Runway 32 at Smyrna/Rutherford County Airport in Tennessee, some nine miles short of their intended destination of Runway 31 at Nashville International, and the crew of an Embraer EMB-120 Brasilia landed at Columbus/Lowndes County Airport, Mississippi, instead of Golden Triangle Regional located about 11 miles away. These incidents happened on clear dark nights in good visibility (ASRS Report No: 478990; 451009).

Black-Hole Approach

Relying on external visual cues at night while conducting approaches can be particularly risky. The sparsity of visual references available to guide one's approach during dark-night conditions has led pilots to conduct excessively low approaches, resulting in fatal crashes short of the runway. As previously explained when discussing dark-night takeoff accidents, black-hole conditions exist on dark nights (moonless and/or overcast sky) when there are few, if any, surface lights in front of the aircraft. If you try to use only outside visual cues to judge your approach angle in these conditions, you may end up too low and risk colliding short of the runway. Conrad Kraft and his colleagues verified this problem in a series of simulator studies conducted in the late 1960s. They concluded that pilots who rely on outside visual cues, without using inside cockpit instruments or outside approach slope indicator aids (e.g., VASI, PAPI), attempt to maintain a constant visual angle between the runway threshold and runway end lights (or the ground lights beyond), which results in a curved approach path, bringing the aircraft dangerously low on the approach.

Ever since Kraft's discovery, the hazards of black-hole approaches have been widely publicized in the aviation community. Unfortunately, pilots are still falling prey to this visual illusion. The experienced flight crew of a Canadian Armed Forces (CAF) Lockheed C-130 Hercules crashed short of Alert Airport, located in the Canadian Arctic, while conducting a visual approach in completely clear weather conditions at night. (This accident and subsequent story of survival was dramatized in the made-for-TV movie *Ordeal in the Arctic*.) More recently, a Federal Express cargo B-727 crashed short of Runway 09 at Tallahassee Regional Airport in Florida; night VMC prevailed. Fortunately the captain, FO and FE survived the accident, but with serious injuries (Report No: NTSB/AAR-04/02). The black-hole phenomenon, along with several other illusions whose power to deceive our senses is strengthened in the dark, will be looked at in greater depth in Chapter 8 on visual illusions.

Even though most of the hazards associated with visual flight at night have been known in the aviation industry for years, the frequency with which these types of accidents still occur indicates that knowledge of these hazards among pilots is not universal. The information has often not trickled down to less-experienced or newly certificated pilots. In a black-hole approach accident in Moosonee, Ontario in Canada, the TSB determined the captain of a Beechcraft C99 Airliner transporting passengers on a scheduled commercial flight hadn't heard of the existence of the illusion and that neither crew member had received any training in aviation physiology or aviation psychology (TSB Report No: A90H0002). Another pilot, in response to a question beforehand about the safety of **night flying** and who that very evening died in a typical dark-night takeoff accident at Grand Canyon National Park Airport in Arizona, stated that flying at night "was no different than flying in the daytime" (NTSB Identification No: LAX96FA052). It may be true that flying at night is no different than flying during the day, except, as the old adage says, *you can't see anything*.

Physiological Limitations of Night Vision

Research indicates that the visual difficulties of piloting an aircraft in the dark arise from both physiological and perceptual limitations inherent in our visual systems. To safely fly at night, you need a basic understanding and thorough appreciation of these limitations. Loss of dark adaptation, monochromatic vision, central vision blindness, reduction in visual acuity and depth perception, night myopia, visual hypoxia, night blindness, and the inevitable effects of aging are just some of the major visual limitations that manifest themselves at night. Some of these limitations were covered in Chapter 2 in the discussion on midair collision avoidance and will be only briefly mentioned here.

Loss of Dark Adaptation

As learned in Chapter 2, the rod receptors in the retina are responsible for true night vision. With higher light intensities the pigment rhodopsin in the rods is broken down (or bleached), and with lower illumination it is replenished. If the rods are even momentarily exposed to high-intensity light levels, then excessive bleaching will occur, rod sensitivity will decrease and it will take additional time (usually 30 minutes) for your eyes to re-adapt (or dark-adapt) to low light levels as the rhodopsin regeneration process begins all over again.

Also, the rods, unlike the cones, are insensitive to light frequencies above 640 nanometers that correspond to the color red. Therefore, if you use red light (or wear red goggles) before flight while in a brightly lit environment, it will enhance your dark adaptation. However, there are problems with the use of red light as the sole source of cockpit illumination. Color discrimination for charts, maps and the instrument panel is severely hampered, and anything colored red (like magenta lines or shading found on air navigation charts) is virtually impossible to see. This, along with the fact that it is difficult to focus in red light (especially for older pilots), is why a low-density white light is generally recommended for interior cockpit lighting. The resulting mesopic vision, where both rods and cones are activated at the same time, is a compromise between preserving dark adaptation and maintaining an adequate degree of color discrimination.

There is actually one situation in which it may be advantageous for you to fly with the interior cockpit lights at full-bright settings: when flying in the vicinity of thunderstorms. The intensity of momentary lightning flashes can completely destroy your dark adaptation, making subsequent vision extremely difficult in the darkened cabin.

Monochromatic Vision

To be able to discriminate colors, two or more types of retinal receptors (and pigments) are required. There are three different types of cones but only one type of rod receptor; hence, the loss of color vision when using pure night (scotopic) vision. Colors, therefore, are seen as various shades of gray.

Central Vision Blindness and Reduction in Visual Acuity

In Chapter 2 you also learned that visual acuity—the ability to visually discriminate fine detail—during daylight conditions is best up to a maximum of only about two degrees either side of the fovea (center of vision) and that there are no rods in the fovea. Therefore, since your foveal cones are shut down during pure night vision, staring directly at a dimly lit object at night will actually make it seem to disappear.

Another reason for lower acuity at night is the high synaptic ratio of rods to optic nerve fibers. While foveal cones have a one-to-one synaptic connection to the optic nerve fibers that carry the visual input to the visual cortex in the brain, there may be as many as 100 to 1,000 rods connected to only a single fiber.[16] The downside is the connection of many rods to only one optic nerve fiber communicates less detail to your brain than the neighboring cones that have a one-to-one connection. The upside is this high ratio contributes to the rods' greater sensitivity to weak light, a fact that enables you—with fully dark-adapted eyes—to see candlelight from as far away as 15 to 20 miles.

Depth Perception Degradation

The high ratio of rods to nerve fibers is also partly responsible for the reduction in your **depth perception** during night vision. Because visual input from many rods funnels into only one neuron to the brain, there is "ambiguity in the exact direction and size of retinal images."[17] Therefore, your judgments of distance are adversely affected.

Dark-Focus (Night Myopia)

In the absence of a target to focus on when looking outside the cockpit during empty-field conditions (high altitude, reduced visibility or at night), your eyes will not focus on optical infinity but will instead focus at their resting state of about two feet to two yards ahead. During the hours of darkness this empty-field myopia—also called dark-focus or night myopia—creates a nearsightedness that blurs objects in the distance, making them appear smaller, farther away and obviously less detectable.

Visual Hypoxia

In Chapter 6, you learned that the use of supplemental oxygen is mandatory in most countries for commercial flights at altitudes above 10,000 feet MSL in unpressurized aircraft. However, its use is also recommended for altitudes above 5,000 feet MSL when flying at night. This recommendation is not made because the composition of night air is any different than day air, but because hypoxia seriously degrades your vision in low-light conditions, partly because of the oxygen requirements of the retinal rod receptors themselves. In fact, the retina has the "highest oxygen demand and the lowest deprivation tolerance of any human structure."[18]

Another consideration is the phenomenon of corneal hypoxia. Since the cornea is removed from the blood supply, oxygen must be directly diffused into it from the outside atmosphere. Therefore, low illumination combined with visual structures that are sensitive to even mild degrees of hypoxia can drastically reduce your visual acuity at night. Carbon monoxide from inhaled cigarette smoke or other sources produces similar effects, since overall oxygen intake from the blood to the retina is reduced. Besides avoiding cigarette smoke, you should also maintain an adequate diet since reduced blood sugar has the same effect on acuity as reduced oxygen.

Night Blindness

On the subject of diet, you should be aware that "eating your carrots" might actually be good for your vision. **Night blindness**—not complete, but functional blindness—can occur during low light conditions. This occurs because of a Vitamin A deficiency in the rod receptors, since rhodopsin is made up of Vitamin A and opsin. Studies have shown that night blindness can be induced within 60 days in normal individuals who have been put on a diet lacking Vitamin A.[19]

Effects of Aging

Time eventually catches up with us all, so it's important you recognize its effects on your night vision. A number of these have been identified. For example, dark adaptation time increases and light sensitivity decreases with age. The lens of the eye actually changes with age; it gets bigger and loses its elasticity, making focusing (accommodation) more difficult. In fact, the near point of vision (the closest distance a person can focus without blurring) will move outward by as much as 10 times for the average 60-year-old. This phenomenon, known as **presbyopia** (old eye), may require the use of corrective lenses. In people over age 40 the far point of focus also tends to move outward, degrading acuity for distant objects as well. Increased light scattering from an aged lens causes older eyes to be more sensitive to glare than younger ones. Because the clear crystalline lens also takes on a bit of a yellow hue with age, reduced sensitivity to the shorter wavelengths of the visible spectrum (blue) occurs. Finally, there is a significant reduction of peripheral vision with age.[20]

Managing the Hazards of Flying in the Dark

Of all the senses we use to pilot our aircraft, we are most dependent on our sense of sight. Unfortunately, as this chapter has made clear, our visual abilities at night are significantly degraded. The accident record indicates that not only does the probability of certain types of accidents increase at night (e.g., dark-night takeoffs; inadvertent VFR flight into IMC or terrain; black-hole approaches), but also that the consequences of these accidents are often fatal. If risk equals the probability of an event times its consequences, then the risk involved in flying using only outside natural visual cues for guidance increases at night. When external visual stimuli are sparse or ambiguous, as is often the case at night, then our ability to accurately perceive the outside world is significantly hampered. An appreciation of the physiological limitations of night vision, along with an understanding of the specific hazards associated with the various phases of flight, is an essential first step in effectively eliminating or reducing their effects. To manage the risks of visual flight at night, it is recommended that pilots supplement natural outside visual references with cockpit flight instruments, navigation instruments and airport aids to vision.

Ground Operations at Night

It's important that you attain and maintain proper dark adaptation before and during a flight at night. Allow sufficient dark-adaptation time before you begin and keep interior cockpit lights low during flight, but not so low as to reduce your ability to accurately read your instruments. Try to avoid exposure to bright lights. If you can't avoid them, then allow only one eye to be exposed; this will at least preserve dark adaptation in the other. Out of courtesy to others, momentarily turn off your taxi lights should they shine into someone else's cockpit. To keep from taxiing too fast, avoid staring straight ahead at the yellow taxiway line or green taxiway centerline lights, and occasionally look out the side window to see more-distant objects (light posts, buildings, etc.) to better judge your speed. Use an airport taxi diagram to help you successfully navigate. Finally, don't hesitate to request progressive taxi instructions if you are unfamiliar with the airport layout. See Chapter 1 for more suggestions on how to safely navigate on the airport movement area at night.

Departing Into the Darkness

To avoid a dark-night takeoff accident when conducting a VFR departure in black-hole conditions, supplement outside visual references with flight instruments such as the airspeed indicator, vertical speed indicator and altimeter to ensure an adequate climb gradient is maintained. Continue to reference them until unambiguous outside visual cues are established. If you have a choice, do not takeoff into black-hole conditions but depart over well-lit terrain.

Flying En Route at Night

If VFR, fly at minimum safe obstruction clearance altitudes and, if possible, over well-lit terrain. Learn how to find the **maximum elevation figure** (MEF) on VFR sectional charts for the quadrant over which you are flying and add at least 1,000 feet (2,000 feet in the mountains) to that figure, since it only provides 100 to 300 feet of obstruction clearance. If IFR, fly at or above the applicable **minimum IFR altitude**. If you don't have an instrument rating, learn how to determine and fly at or above minimum en route altitudes as shown on IFR charts and, if off an airway, at minimum off-route obstruction clearance altitudes. Make sure you know your position at all times and use radio navigation aids, and ATC assistance if necessary, to avoid geographic disorientation. Many aircraft, including smaller GA aircraft, are now equipped with **terrain awareness and warning systems** (TAWS). These systems, which co-locate your GPS position with a terrain and obstacle database, can help you maintain your positional awareness and provide you with terrain avoidance information as well. But don't become complacent and lose your ability to navigate using other tried-and-true methods. The use of TAWS and enhanced GPWS to avoid a CFIT accident are more fully explained in Chapter 10.

As discussed in Chapter 4, your personal VFR weather minimums should be higher at night than they are in the day. However, they mean nothing if you are not resolved to stick to them when weather conditions drop below them. Pay particular attention to visibility and cloud reports and forecasts, and be aware that you can't always count on your vision to avoid inadvertent entry into IMC. If flying NVFR, it's best to fly in **high lighting conditions**, which the FAA *AIM* defines as either flight over surface lighting that provides for the lighting of prominent obstacles, the identification of terrain features (shorelines, valleys, hills, mountains, slopes) and a horizontal reference by which you can control the aircraft, or a sky condition with less than broken (5/8) cloud coverage and a moon with at least 50 percent illumination. This lighting could be a result of extensive cultural (human-made) lighting found over metropolitan areas or limited cultural lighting combined with a high level of natural reflectivity of celestial illumination, such as that provided by a snow-covered surface or a desert surface. Finally, if you don't have extensive mountain flying experience and knowledge of local terrain, you should completely avoid flying VFR over mountainous terrain at night.

Approach and Landing at Night

Some airports are more prone to illusions at night than others. Therefore, consult the official airport publications—the *A/FD* (or the *CFS* in Canada)—and other applicable sources (e.g., airport operators, other pilots) to determine which airports are more conducive to visual illusions at night.

When approaching a nontowered airport at night, especially one you are unfamiliar with, consider overflying it beforehand to ascertain the presence of any terrain or obstruction hazards. Avoid long straight-in visual approaches; premature descents on straight-in approaches are common even in the daylight. Use altitude and glide path instruments (e.g., ILS; distance measuring equipment [DME] distances correlated with altimeter readouts; GPS guidance) to assure safe obstacle clearance while conducting an approach. A constant descent of about 300 feet per NM results in a 3-degree glide slope. A descent of 500 feet per NM approximates a 5-degree approach angle—which may be preferable to ensure adequate obstacle clearance at airports you are unfamiliar with. Understand and use **visual approach slope indicator systems** such as VASI and PAPI while on the approach. While many of these systems are visible up to 20 miles or more at night, according to the FAA *AIM*, the visual glide path generally provides safe obstruction clearance only up to plus or minus 10 degrees of the extended runway centerline (6 to 9 degrees on either side in Canada) to a distance of only 4 NM from the runway threshold. The NTSB determined that the PAPI for Runway 09 at Tallahassee was fully functional when the Federal Express pilots allowed their B-727 to strike the trees short of the runway; however, the crew failed to effectively use the system.

Additional strategies on how to conduct safe approaches—during the day or night—are outlined in Chapter 8 on how to effectively cope with visual illusions, and in Chapter 10 on how to avoid a CFIT accident.

In spite of the dual visual receptor system designed to help us adapt to reduced lighting conditions at night—cones for day and rods for night—our ability to accurately perceive the outside world can be severely impaired. We should not completely trust our senses; sometimes they deceive us. If we recognize these limitations and are aware of the flight situations in which we are likely to be vulnerable, flying at night can not only be a pleasant experience but an accident-free one as well.

Helpful Resources

"Darkness Increases Risks of Flight" is an article written by Dale Wilson documenting the increased risks of nighttime flying and it provides a foundation for much of this chapter. It was published in the November/December 1999 issue of *Human Factors & Aviation Medicine*, produced by the Flight Safety Foundation. (FlightSafety.org/archives-and-resources/publications/human-factors-aviation-medicine)

Controlled Flight Into Terrain in Visual Conditions: Nighttime Visual Flight Operations Are Resulting in Avoidable Accidents is a 2008 NTSB *Safety Alert* (SA-013) that lists some notable accidents in which aircraft were flown into terrain in VFR weather conditions at night, and summarizes countermeasures you can take to prevent such an accident from happening to you. (www.NTSB.gov/safety/safety_alerts.html)

Several articles on night flying were published in the FAA's magazine for GA safety, the *FAA Safety Briefing* (formally the *FAA Aviation News*). (www.FAA.gov/news/safety_briefing)
- *Shedding Light on Night Flight*, November/December 2008
- *N.I.G.H.T.*, November/December 2005
- *Be Aware of What Lurks in the Night*, November/December 2003

The AOPA Air Safety Institute provides several resources on night flying, including safety articles, briefings, advisors, quizzes and videos. (www.AOPA.org/Pilot-Resources/Safety-and-Technique/Operations/Night-Flying.aspx)

Transport Canada produces a *Night VFR Prevention Tools* CD-ROM that includes several articles (including one written by one of this book's authors), accident reports, slides and posters. Many of these can be accessed individually online for free or are available as part of the six-CD set System Safety Summer Briefing Kit (TP 14112), which can be ordered online (for a fee). (www.TC.gc.ca/eng/civilaviation/publications/tp14112-menu-316.htm)

Ordeal in the Arctic is a made-for-TV movie dramatizing the crash of the Canadian Armed Forces Lockheed C-130 Hercules (Flight *Boxtop 22*) near Alert in the Canadian territory of Nunavut (formerly the Northwest Territories) after the crew succumbed to the black-hole illusion, and the subsequent survival and rescue of the 14 of 18 people onboard who survived. Released by Trimark Home Video in VHS in 1997 and DVD in 2002. Check online retailers for availability.

Notes

1. National Safety Council, *Driving at Night* (Itasca, IL: NSC, 2009). Available online at www.NSC. org/news_resources/Resources/Pages/SafetyHealthFactSheets.aspx.

2. AOPA Air Safety Institute, *2008 Nall Report: Accident Trends and Factors for 2007* (Frederick, MD: AOPA ASI, 2009).

3. National Transportation Safety Board, *NTSB Safety Alert, Controlled Flight Into Terrain in Visual Conditions: Nighttime Visual Flight Operations Are Resulting in Avoidable Accidents*, SA-013 (Washington, DC: NTSB, January 2008). Available online at www.NTSB.gov/safety/safety_alerts. html.

4. Dale R. Wilson, "Darkness Increases Risks of Flight," *Human Factors & Aviation Medicine* 46 (November–December 1999): 1–8.

5. Roy W. Chamberlin, "Rejected Takeoffs: Causes, Problems and Consequences," *Flight Safety Digest* 12 (January 1993): 1–7.

6. "Pitch Up, Pitch Down, Pitch Black," *Aviation Safety Reflexions* 4 (December 1993): 1–4.

7. AOPA Air Safety Institute, *Safety Review, General Aviation Weather Accidents: An Analysis and Preventive Strategies* (Frederick, MD: AOPA ASI, 1996).

8. T.P. Miller and T.J. Casadevall, "Volcanic Ash Hazards to Aviation," *Encyclopedia of Volcanoes*, ed. H. Sigurdsson (Academic Press: San Diego, 2000): 915–930.

9. Boeing Commercial Airplanes, *Statistical Summary of Commercial Jet Airplane Accidents: Worldwide Operations, 1959–2010* (Seattle, WA: Boeing, June 2011).

10. Ronald Ashford, "A Study of Fatal Approach-and-Landing Accidents Worldwide, 1980–1996," *Flight Safety Digest* 17 (February–March 1998): 1–41.

11. Scott A. Shappell and Douglas A. Wiegmann, "U.S. Naval Aviation Mishaps, 1977–92: Differences Between Single- and Dual-Piloted Aircraft," *Aviation, Space, and Environmental Medicine* 67 (1996): 65–69.

12. R. Khatwa and A.L.C. Roelen, "An Analysis of Controlled-Flight-Into-Terrain (CFIT) Accidents of Commercial Operators, 1988 Through 1994," *Flight Safety Digest* 15 (April–May 1996): 1–45.

13. Melchor J. Antunano and Stanley R. Mohler, "Geographic Disorientation: Approaching and Landing at the Wrong Airport," *Aviation, Space, and Environmental Medicine* 60 (1989): 996–1004.

14. "Convair Hits Trees at Site Mistaken for Airport," *Flight Safety Digest* 12 (March 1993): 23.

15. David Hughes, "Safety Group Highlights CFIT Risk for Regionals," *Aviation Week & Space Technology* 140 (May 9, 1994): 46–51.

16. Warner D. Fan, "Night Vision: Question and Answer Column," *Aviation, Space, and Environmental Medicine* 62 (March 1991): 274–275.

17. Ibid.

18. David A. Tipton, "A Review of Vision Physiology," *Aviation, Space, and Environmental Medicine* 55 (1984): 145–149.

19. S. Coren, C. Porac and L.M. Ward, *Sensation and Perception*, 2nd ed. (Orlando. FL: Academic Press, 1984): 152.

20. D.H. McBurney and V.B. Collings, *Introduction to Sensation/Perception*, 2nd ed. (Englewood Cliffs: Prentice-Hall, 1984).

What You See Is Not Always What You Get

8

A student pilot flying a Cessna notices a well-manicured grass airstrip below him and decides to come in for a landing. After sliding off the end of this 200-foot runway, he's astonished to discover that he just tried to land at a miniature airport built for radio-controlled (RC) aircraft.[1] A commercial pilot and five passengers are on a Part 135 VFR sightseeing flight when their helicopter crashes on the Juneau Icefield in Alaska. The weather is marginal VFR with localized snow showers on this overcast day, but the pilot thought the aircraft was clear of the surface by at least 500 feet.[2]

These pilots had two things in common: They were both deceived by their senses, and unlike many other pilots who've been similarly duped, they were able to walk away (although a passenger aboard the helicopter received serious injuries). The novice pilot thought the runway looked fine, and the commercial helicopter pilot, while flying in **flat light** conditions over a large, featureless snow-covered ice field, was sure he had plenty of altitude. They were both victims of what perception scientists call a **visual illusion**. A visual illusion is a misperception of visual reality; in other words, what you see is not what you get.

You have probably seen a variety of geometric illusions such as the Ponzo, Poggendorf and Muller-Lyer illusions (Figure 8-1). There are more than 200 of these, most of them discovered and cataloged by physiologists and artists almost a century ago. What they have in common is their ability to trick you into seeing something that isn't there. Nobody knows exactly why they do this; in fact there are many theories, and even categories of theories, to explain them. What we do know is that pilots are not immune from visual illusions, and even though they occur relatively infrequently in flight, when they do the consequences are often fatal. That's because by definition visual illusions *deceive* you into thinking everything looks fine and you are therefore in no danger.

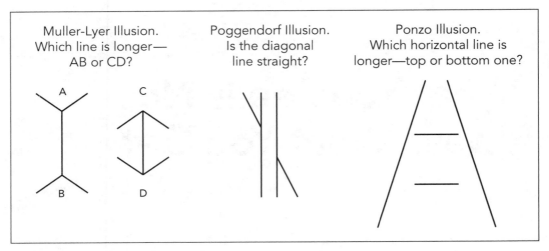

Figure 8-1. Examples of geometric illusions.

Considerable research has been conducted over the past few decades, and a number of visual illusions pertinent to the flight environment have been identified. The previous chapter introduced this topic by discussing a particularly deadly illusion that occurs on clear dark nights: the **black-hole approach illusion**. There are several others. Most occur during the approach and landing phase, but some can fool pilots during any phase of flight. A few occur during the day in good light conditions, but most occur in conditions of limited visual cues such as at night or in reduced visibility.

Approach and Landing Illusions

As mentioned in Chapter 7, more than a third of the world's fatal commercial jet airplane accidents occur during the final approach and landing phases of flight. Because misinterpretation of outside visual cues has been responsible for several of these accidents, the primary focus of this chapter is on **visual landing illusions**. Pilots have unexpectedly crashed short of the runway or landed too hard on it, while others have landed long and run out of stopping distance. Some illusions during approach and landing are caused by certain atmospheric and runway lighting conditions; the majority, however, involve sloped runways, sloped terrain, runways with dimensions you are not used to, and approaches and landings on runways in the dark.

Sloping Runways

We usually judge our approach angle by the familiar trapezoidal image the runway shape casts upon our retina (Figure 8-2). However, when flying on the correct approach angle to a sloped runway, depending on the direction of runway slope, a high-approach runway image or low-approach runway image is projected onto the retina of each eye. When you are on the proper approach angle to a downsloping runway, for example, the retinal image will be a low-approach shape, giving you the illusion you are too low. Figure 8-3 illustrates that in response to this false perception you will instinctively adjust your approach angle to visually acquire the normal approach image you are accustomed to; the resulting high approach could lead you to a go-around at best or a long landing and runway overrun at worst.

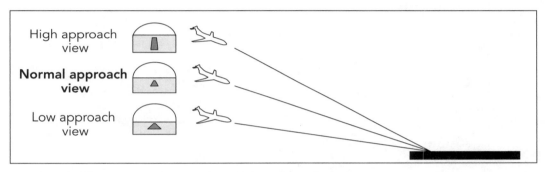

Figure 8-2. Different runway approach views from the cockpit.

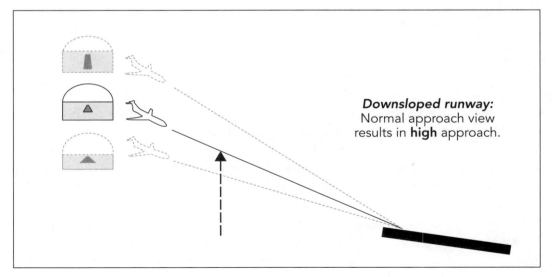

Figure 8-3. Effect of downsloped runway on approach.

Figure 8-4 indicates that a high-approach image on your retina during a proper approach to an upsloped runway creates the illusion that you are too high, leading you to conduct a low approach. This illusion is deadly; the resulting low approach could lead to a premature hard landing at best or a CFIT accident short of the runway at worst. Many pilots—novice and expert alike—have fallen prey to this illusion. This can happen during the day but is considerably strengthened at night, especially on a dark night. For example, the experienced crew of a Canadian Airlines International Boeing 767 was fooled by the upsloping runway at Halifax Stanfield International Airport in Nova Scotia during a nonprecision approach to Runway 06 at night. The 0.77 percent upslope tricked both pilots into thinking the airplane was too high, and in spite of mostly red PAPI lights (indicating it was too low), they flew a low approach on short final that resulted in a premature hard landing and tail strike (TSB Report No: A96A0035).

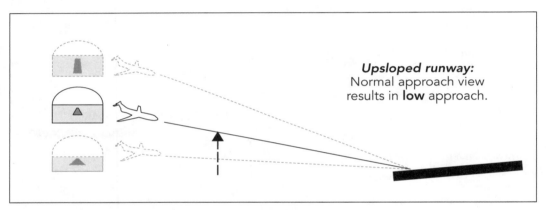

Figure 8-4. Effect of upsloped runway on approach.

These illusions are compounded when the runway is irregularly sloped. For example, a visual loss of the runway end on an upsloping runway whose latter portion is level has led to blown tires because of needless braking or collision with trees from attempting an impossible go-around. For example, the unevenly sloped runway at Catalina Airport in Avalon, California, has contributed to several accidents and incidents. The first 2,000 feet of the 3,240-foot-long runway has a pronounced upslope gradient, and according to the *A/FD*, pilots cannot see aircraft at opposite ends of the runway due to its gradient. The pilot of a Cessna 172 had trouble conducting a proper flare-out and ended up bouncing, porpoising, and losing directional control on Runway 22. He thought he was nearing the end of the runway, so he intentionally turned left to avoid running off what he thought was the departure end. He collided with three parked aircraft, but luckily, no one was hurt (NTSB Identification No: LAX96LA235).

Sloping Terrain

Sloping terrain underneath your landing approach path can also fool you into thinking you are too high or too low on the approach. For example, terrain that slopes down toward the threshold can create the illusion that you are *too low* (especially in good daylight conditions), leading to a high approach and possible long landing or go-around (Figure 8-5). Even though the retinal image indicates that your aircraft is on the correct approach angle, peripheral vision and the role of what perception scientists call optic flow is dominant and overrides the normal approach image on your retina. The phenomenon of optic flow was introduced in Chapter 1. It is the rate at which objects in our peripheral vision flow past us as we move through space, and its value is a ratio of speed divided by altitude.[3] For example, if you've ever experienced the difference between driving a close-to-the-ground sports car and a big semitrailer truck at the same speed, then you understand this concept. The speed of the car seems much faster because of the increased optic flow in your peripheral vision. In fact, a sports car traveling 40 mph has a higher optic flow rate than a jet traveling 400 mph at 40,000 feet. As mentioned in Chapter 1, this explains why early B-747 pilots, sitting at a significantly higher eye-to-wheel height than they were previously accustomed to, tended to taxi too fast and why all pilots tend to do the same at night.

Terrain that slopes up toward the runway threshold potentially poses a greater hazard. The significantly reduced optic flow in your peripheral vision while conducting an approach creates the illusion the aircraft is higher than it actually is, enticing you to fly lower than normal, which could result in a possible CFIT accident short of the runway (Figure 8-5). Mountain pilots often have to contend with both upsloping runways and upsloping terrain at the same time. The combination of the high-approach retinal image from the upsloping runway and the reduced optic flow from the upsloping terrain below their flight path creates a very strong height illusion that can lead to a dangerously low approach. Both of these phenomena were operative in the Catalina Airport **runway excursion** accident. The airport sits atop a mountain with steeply upsloping terrain at each end of the runway; so not only is the first half of the runway sloped upward, but the terrain leading up to it is as well.

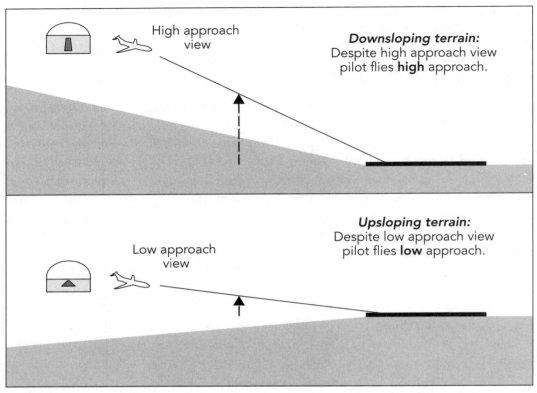

Figure 8-5. Effect of downsloping and upsloping terrain on approach. The influence of optic flow in a pilot's peripheral vision is dominant in these illusions.

Runway Dimensions

If an object of known size (e.g., automobile, adult human, airplane) casts a smaller image on your retina, you rightly perceive it to be farther away, not shrunken in size. Perception scientists call this the **relative size cue** to distance perception. Problems occur, however, when we think the size hasn't changed when in fact it has. For example, compared to larger North American-built automobiles, a greater proportion of smaller foreign-manufactured cars were rear-ended by other vehicles when they were first introduced to the U.S. and Canadian markets. Drivers incorrectly perceived the smaller (narrower) cars as larger ones that were farther away than in reality. As a result of this illusion, drivers who were accustomed to seeing only larger cars in front of them tended to apply their brakes too late, resulting in rear-end collisions.[4]

As the student pilot who landed his Cessna on the RC strip discovered, the same illusion occurs to pilots who think a smaller runway is the same size as their home runway, but farther away in distance. The illusion is particularly strong when the unfamiliar runway has the same length-to-width (L/W) ratio of the runway to which you are accustomed. For example, the larger runway in Figure 8-6 is 5,000 feet long by 200 feet wide, and the smaller runway 2,500 feet by 100 feet. Both have the same proportions (L/W=25), making it difficult to tell them apart—especially when there is little or no surrounding context to provide cues to their actual size, such as in conditions of low-visibility or during the hours of darkness.

The explanation of this illusion is that pilots who experience it suffer from a malady popularly known as the **home-drome syndrome**. Your home-base aerodrome (airport) is so indelibly etched into your perceptual memory that it's easy to overestimate the distance of a runway that is smaller than the one you're accustomed to, and underestimate the distance of a runway larger than you're accustomed to, especially if you are a low-time pilot who lacks experience with different-sized runways.

Smaller-Than-Usual Runways

When approaching a runway that is smaller than what you are used to (the runway on the right in Figure 8-6), the relative size cue creates the illusion of increased distance between the runway and your aircraft. This illusion can lead to either a high approach or a low approach, depending on whether you perceive the runway as farther away horizontally or vertically. The NTSB cited the visual illusion of a smaller and narrower runway as partially responsible for the low approach of a Piper Cheyenne on Runway 06 during civil twilight conditions at Chicago Executive Airport in Wheeling, Illinois. The PA-31T struck a 25-foot high unlighted utility pole 750 feet from the landing threshold and was able to land, but veered off the runway due to landing gear damage. Fortunately, the two commercial pilots aboard were uninjured (NTSB Identification No: CHI03IA108).

Regardless of how you might perceive this illusion of increased distance when conducting an approach to a runway that is smaller than you are used to, the illusion will also alter your perception of when you should initiate the landing flare. The smaller runway will appear to be farther away in your visual field, so you are likely to initiate the flare later than usual, resulting in a hard landing.

Larger-Than-Usual Runways

Conversely, when approaching a runway that is larger than what you are familiar with (the runway on the left in Figure 8-6), the relative size cue creates the illusion of decreased distance between your aircraft and the runway. Depending on whether you interpret this as vertical distance or horizontal distance, you might respond by conducting an approach that

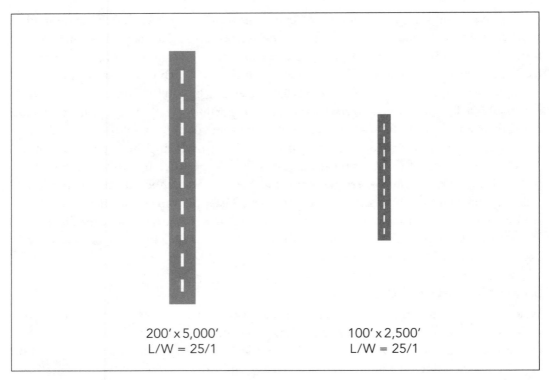

200' x 5,000'
L/W = 25/1

100' x 2,500'
L/W = 25/1

Figure 8-6. Different-sized runways with same L/W ratio.

is either too high—resulting in a possible long landing or go-around—or too low—resulting in possible ground impact short of the runway. If you manage to overcome this illusion and make it safely to the runway threshold, the blossoming of the runway environment in your peripheral vision will often lead to a high flare-out and a possible stall over the runway. This is what happened to a pilot who landed hard and bounced several times before the propeller of his Cessna 172 struck the runway at Asheville Regional Airport in North Carolina. The pilot reported the runway was twice as long and twice as wide as his home runway, causing him to initiate the landing flare too high above it (NTSB Identification No: NYC07CA116).

The home-drome syndrome also can create strong illusions when approaching runways with different L/W ratios than what you are accustomed to (Figure 8-7). For example, research in the 1980s confirmed that a low-approach illusion occurs when conducting an approach to a runway with a smaller L/W ratio (the runway on the left in Figure 8-7) than what you are used to, leading you to fly a high approach. The L/W ratio decreases with wider and/or shorter runways.[5] A more critical situation arises when the L/W ratio is greater than what you are used to. With a longer and/or narrower runway (the runway on the right in

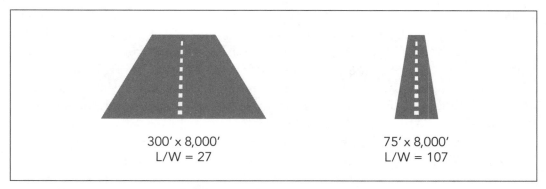

300' x 8,000'
L/W = 27

75' x 8,000'
L/W = 107

Figure 8-7. Runways with different L/W ratios.[6]

Figure 8-7), the illusion of being too high may induce you to fly an unsafe low approach, resulting in a premature hard landing or CFIT accident short of the runway.

A student pilot flying a Piper PA-28 was killed when her aircraft struck the top of a moving semitrailer truck as she conducted an approach to Runway 26L at the Sylvania Airport in Sturtevant, Wisconsin. Witnesses reported that the Cherokee approached the runway at a "very low angle" as it neared the four-lane highway that runs perpendicular to the final approach path. The pilot was used to landing at her home airport on runways that were 100-feet and 150-feet wide, which corresponded to L/W ratios of 32 and 28, respectively. The accident runway at this different airport was only 33 feet wide and had an L/W of 70—more than twice what she was used to. The NTSB cited this as a causal factor in the accident (NTSB Identification No: CHI98LA061).

Atmospheric Conditions

Objects in the distance appear darker, bluer and more blurred due to a greater degree of light scattering from molecules and particulate matter between the object and observer. Closer objects appear brighter, with finer detail and truer color, due to less light scattering. Therefore, perceived distances of objects not only depend on the relative size but also on the opacity of the atmosphere, a phenomenon discussed in Chapter 2 in the context of identifying traffic conflicts, known as aerial (atmoshperic) perspective.[7] On exceptionally hazy days you might perceive an object—including a runway—as farther away than it really is. This could result in a higher or lower approach depending on how your brain perceives this illusion of distance: If you interpret this illusion vertically as an increase in altitude, you could be tempted to begin the approach prematurely and conduct a low approach; if you perceive it primarily as an increase in horizontal distance from the runway, you might commence the approach later than normal, which would put you too high on the approach.

On unusually clear days, which often occur during cold dry winter conditions, the opposite illusion can occur: objects appear closer than they actually are. If you perceive this as less horizontal distance between your aircraft and the runway, you could end up beginning your descent too far from the airport, resulting in a dangerously low approach. The captain of the CAF Lockheed C-130 Hercules that crashed short of the runway at Canadian Forces Station Alert in black-hole conditions at night (mentioned in Chapter 7) also was fooled into thinking his aircraft was closer to the airport than it actually was, due to extremely clear air conditions. This contributed to his premature descent and the airplane's contact with terrain about 18 miles short of the airport.

Runway Lighting

The intensity of runway lights can affect your ability to conduct a successful approach at night. This **relative brightness cue** arises out of the atmospheric effects noted above: Since objects farther away are darker (aerial perspective cue) and smaller (relative size cue), darker objects that are close are sometimes erroneously perceived to be farther away and smaller than they really are. The opposite is also true: Since closer objects are brighter and larger than objects farther away, brighter objects are sometimes perceived to be closer and larger than they are in actuality. For example, brighter-than-normal runway lights create the illusion of being closer, while dimmer-than-normal runway lights give the illusion of being farther away. As we've discussed, this can lead to different responses depending on whether your brain interprets this distance illusion vertically or horizontally. One thing is certain, however: Dimmer runway lights often lead to late flare-outs and hard landings, while brighter lights usually have the opposite effect.

Duck-Under Phenomenon

After conducting an approach to IFR landing minimums, some pilots have found themselves inadvertently descending below the glide path after transitioning to visual references. One possible explanation of this **duck-under phenomenon** is as follows: With good visibility and higher ceilings, you are accustomed to seeing the far end of the runway a certain distance up from the panel on the windscreen (Figure 8-8). However, upon reaching decision altitude/height at IFR visibility minimums (usually 1/2 SM or runway visual range [RVR] of 2,600 feet), the far end of the runway now appears to be lower on the windscreen, causing you to mistakenly believe the aircraft (and its nose) is higher than it should be. This

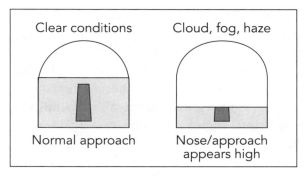

Figure 8-8. Duck-under phenomenon.

could induce you to lower the nose, causing an excessively low approach and a possible hard landing short of the runway.

The Black-Hole Approach

Undoubtedly the most hazardous approach and landing illusion is the black-hole approach illusion. As mentioned in the previous chapter, black-hole conditions exist on dark nights (usually with no moon or starlight) with unlighted terrain between the aircraft and the runway environment. These conditions have resulted in pilots conducting dangerously low approaches that for many have ended in fatal CFIT accidents short of the runway.

For example, the black-hole illusion was a factor in a fatal accident involving a Rockwell Sabreliner 65 corporate jet on the approach to Kaunakakai Airport on the island of Molokai in Hawaii. The Sabreliner collided with mountainous terrain 3.3 NM from the airport, killing all six people on board, after the crew terminated the instrument approach and attempted to fly a visual approach in VMC at night (NTSB Identification No: LAX00FA191). According to the NTSB, the black-hole illusion also claimed the lives of two passengers onboard a Part 135 scheduled passenger flight from St. Croix to the Cyril E. King Airport on St. Thomas, in the U.S. Virgin Islands. While the airline transport pilot was conducting a visual approach at night in VMC, the Cessna 402 crashed into the dark waters of the Caribbean Sea about three miles southwest of the airport (NTSB Identification No: MIA97FA082).

In relativity theory, a black hole is a region of space that has such a strong gravitational pull that nothing can escape its grasp. The same could be said about black-hole conditions on an approach to landing. If you are unprepared for it, you will be unable to keep your aircraft from being prematurely pulled toward the earth. Perception scientists don't know for certain what causes this illusion, and it is likely no one theory fully explains it. In Chapter 7, we briefly introduced the research that Boeing scientist Conrad Kraft and his colleagues conducted on this phenomenon. Experienced Boeing instructor pilots (with more than 10,000 hours each) flew entirely visual approaches—with no altimeter or other glide path information available—to runways in black-hole conditions in a simulator. The result? Most of them flew excessively low approaches and crashed into the terrain short of the runway. The researchers explained that in the absence of lighted terrain between the aircraft and runway, pilots attempt to maintain a constant visual angle between the runway threshold and runway end lights (or the ground lights beyond).[8] Contrary to what you might think, a constant *visual* angle does not equal a constant *approach* angle. In fact, as Figure 8-9 illustrates, a constant approach angle results in an ever-increasing visual angle as you get closer to the runway. When you attempt to maintain a constant visual angle, the result is a curved flight path that extends below a safe approach angle (Figure 8-10).

Some researchers claim the original Boeing conclusions are contradictory and that the mechanisms involved in visually landing an aircraft are still not fully understood.[9] Others have suggested that pilots are enticed into flying a low approach because of the visual expansion of the runway environment that occurs when their vision transitions from near-focus

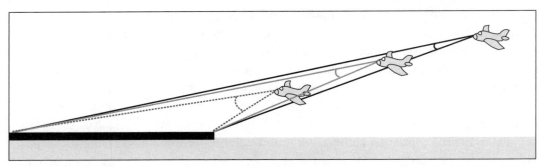

Figure 8-9. Constant approach angle equals increasing visual angle.

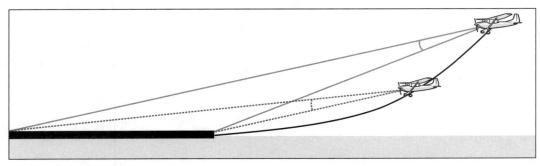

Figure 8-10. Constant visual angle results in curved approach.

(the cockpit instruments) to far-focus (the runway environment) when approaching the airport. This in turn causes an illusion of increased height, which results in a lower approach.[10] Of course, as discussed earlier, a critical visual cue pilots rely on for height perception while conducting approaches during the daylight—optic flow—is completely absent in black-hole conditions. Without this relative motion of outside terrain in their peripheral vision, it's virtually impossible to judge the aircraft's height above the ground. It is likely, therefore, that all of these factors play a role in deceiving pilots into thinking that they are too high when conducting approaches in black-hole conditions.

You should be aware that an upsloping runway only strengthens the black-hole approach illusion. As discussed earlier, an upsloping runway creates the illusion of being too high, causing you to fly a low approach. Add the tendency to descend below the glide path in black-hole conditions and premature flight into terrain short of the runway is likely. In one of Kraft's simulator experiments, 12 senior Boeing instructor pilots flew visual approaches in black-hole conditions to an upsloping runway environment without the aid of an ILS or altimeter: 11 of them crashed short (by a few miles) of the runway![11] More recently, an upsloping runway was implicated in a fatal CFIT accident at the Aspin/Pitkin County Airport in Colorado. The crew of a Grumman Gulfstream III was conducting a visual approach in the dark, and tragically all three crew members and the 15 passengers aboard this commercial charter flight died when it crashed 2,400 feet short of Runway 15. The NTSB stated the upward slope of this runway probably aggravated the black-hole illusion.[12]

In Chapter 2 we stressed the importance of sitting at the correct design eye position (DEP) for optimal visual detection of outside traffic. If your seat is adjusted lower than the DEP while conducting an approach, you might also inadvertently lower the nose and descend below the glide path in an attempt to maintain sight of the runway environment. This happened to the captain of the scheduled Frontier Air flight into Moosonee, Ontario, that we mentioned in Chapter 7. His seat position resulted in his eyes being 4.7 inches lower than the DEP. The TSB report concluded the captain flying the Beechcraft C99 Airliner would not have been able to see anything below the horizon without lowering the nose of the aircraft. Unfortunately, in an attempt to maintain visual contact with the runway lights while conducting a visual approach at night, he pitched the nose too low, which resulted in an excessive descent rate of 1,200 feet per minute. The airplane crashed seven miles short of the runway, killing the FO and seriously injuring the captain and two passengers (Report No: A90H0002).

Unfortunately, awareness and experience with this particular illusion doesn't guarantee immunity from its effects. The flight crews in both the Hawaii and St. Thomas black-hole accidents mentioned earlier were highly experienced pilots with more than 12,000 hours each. In another accident, a certificated flight instructor (CFI), who knew the hazards of the black-hole approach and explained this phenomenon to his student while en route to the airport, still allowed the aircraft to strike the trees on final approach.[13]

Other Illusions in Flight

Most visual illusions occur during the approach and landing, and most of those occur at night. However, your eyes can fool you during other phases of flight—day or night. The following are some of the more well-known that have played tricks on pilots.

Rain on Windscreen

Heavy rain on an aircraft's windscreen can create the illusion of increased height, which in turn could cause you to fly an excessively low approach. Much like the apparent bending of a pencil when placed in a glass of water, the presence of significant amounts of water on your aircraft's windows can produce a prism effect, causing the outside image to appear lower than it really is. The illusion is compounded by the darkening of the image due to the diffusion of light through the water. This relative brightness cue only strengthens the false impression that you are farther away and higher than you actually are. Transport Canada's *AIM* states the downward bending can be as much as 5 degrees, causing an object (e.g., hilltop, runway) one mile away to appear as much as 500 feet lower than it actually is.

Flat Light and Glassy Water

A condition similar to the black-hole phenomenon that occurs during the day is the phenomenon of flat light. Also known as sector whiteout or partial whiteout, flat light is not a zero-zero condition of snow and/or blowing snow (a condition that is usually just called "whiteout"); it is a situation where the aircraft is usually flying above snow-covered featureless terrain below an overcast cloud layer. The visibility could be unlimited; however, the scattering of light and lack of shadow drastically reduces contrast and depth/distance perception. Skiers have experienced this when attempting to traverse snow-covered terrain. For pilots, both the enroute and approach-to-landing phases of flight in such conditions are difficult if not impossible to fly without reference to instruments since, for all practical purposes, no outside visual reference remains.

This phenomenon was one of the major contributors to New Zealand's worst aviation accident to date: The 1979 crash of an Air New Zealand McDonnell Douglas DC-10 into Mount Erebus while on a sightseeing flight in the Antarctic. The crew was unaware the airplane was flying directly toward the rising terrain of the Mount Erebus volcano, in part because the snow-covered surface and overcast sky blended imperceptibly together in sector whiteout conditions. Tragically, all 257 people on board perished.[14] More recently, a helicopter on a sightseeing flight crashed on the Herbert Glacier located on the Juneau Icefield in Alaska, killing the commercial pilot and six passengers. Three months later on that same icefield, three helicopters from the same charter company crashed within two miles of each other while flying over snow-covered featureless terrain: A Eurocopter AS350 crashed while on a sightseeing flight, injuring the commercial pilot and all five passengers; the second Eurocopter crashed in the same area almost three hours later while conducting a search-and-rescue mission for the first helicopter; and two hours later the third helicopter also crashed nearby while looking for the first aircraft. Fortunately, no one died in these three accidents. Overcast cloud and flat light conditions prevailed and were cited by the NTSB as major causal factors.[15]

Reliance on aircraft flight instruments is the only way to accurately determine your aircraft's altitude and attitude in flat light conditions. Unfortunately, only one additional reference is added to your visual field during an approach and landing in these conditions: the runway. Certainly this is an improvement. However, as is the case with black-hole approaches, depth perception while conducting an approach in this environment usually leads to an excessively low approach, increasing the chance of premature contact with the ground.

You should note that loss of depth perception can occur when conducting approaches over large bodies of water, especially calm **glassy water**; it's easy to descend into it without knowing you're about to do so. For example, a pilot was killed and the float-equipped Cessna 172 he was flying was destroyed when it impacted the water during a landing on Ash Lake, near Orr, Minnesota. A pilot witness on the ground reported the airplane did not flare prior to impacting the glassy smooth water. The NTSB cited the visual illusion "caused by the glassy smooth water condition" as a factor in the accident (NTSB Identification No: CHI95FA307).

Various Night-Light Illusions

Many unusual attitudes have resulted at night from pilots who have mistakenly identified ground lights as stars, and stars as ground lights. In a very dark environment, if you stare directly at a stationary light it will appear to move about your field of view. Known as the **autokinetic effect**, this phenomenon can lead to potentially fatal consequences. Pilots have been known to take evasive action to avoid colliding with what they perceived to be another aircraft when in fact it was the light of a lookout tower located several miles away!

The retinal fatigue that results from staring at a sole light source for any appreciable length of time is believed to be the cause of this apparent motion.

Flicker Vertigo

Another phenomenon associated with light is **flicker vertigo**. Sunlight passing through an airplane's propeller or the rotor blades of a helicopter at certain frequencies can create a flicker or strobe-light effect. For example, conducting an approach into the sun at idle power settings (low propeller RPM) can induce this. Even strobe and beacon lights (especially in cloud) can create flicker. The reaction to this ranges from simple distraction, to illusions of movement in the opposite direction of the shadows, nausea, and for some individuals even epileptic-type seizures.

Illusions on the Apron

We've already explained how an illusion of reduced speed occurs when taxiing on the apron at night, which can cause pilots to taxi too fast. Other illusions also can confuse you during ground operations at the airport. Laterally blowing snow on the runway or airport apron can create the illusion of movement when the aircraft is actually stopped, or the impression of being stationary when the aircraft is in fact moving. The movement of passenger loading bridges away from or toward an aircraft can also give you the illusion of aircraft movement, possibly leading to unnecessary braking action.

Overcoming Visual Illusions

Visual illusions are relatively rare events in flight, but as we've seen, when they do occur they can be deadly. They are not usually the result of simple physiological limitations inherent in the human eye; the image on the retina doesn't lie, but our perceptual interpretation of it does. This is especially true in conditions of **ambiguous visual stimuli**. For example, the lines of the classic Necker cube in Figure 8-11 are received accurately by our visual sense, but our perception of them changes. Even though the lines are two-dimensional, the subconscious interpretive part of our brain wants to see them as parts of a three-dimensional cube. One moment the top right two-dimensional square looks like it might be the front of a three-dimensional box, the next moment it appears to be the back of it. This is an example of an ambiguous figure, in which there isn't enough contextual information (stimuli) for us to accurately perceive its orientation. The figure, therefore, keeps reversing. It's almost like

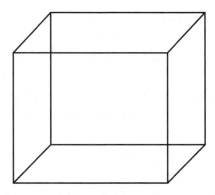

Figure 8-11. Necker cube. An example of an ambiguous figure.

our brain makes two hypotheses (albeit unconsciously) as to which way the cube should be perceived, and since there isn't enough information, it alternates between both of them. When we rely solely on outside visual stimuli to fly our aircraft at night, in poor visibility, or both, we are flying in conditions of visual ambiguity and are therefore vulnerable to visual illusions.

Though the exact mechanisms involved aren't fully understood, visual illusions are real and have tricked many pilots into seeing the outside world incorrectly. Under the right conditions visual illusions can be strong, persistent and deadly; however, you aren't helpless when it comes to coping with them. Being aware of their existence and knowing the situations in which you are most likely to encounter them are important first steps in overcoming them. However, simple awareness of these illusions isn't enough to fully protect you from being deceived. As mentioned in the black-hole discussion, the CFI in the tree-strike incident was fully aware of this illusion at the airport he and his student were approaching—that is why *he* elected to conduct the approach himself. He later reported it was ironic that he was consciously aware of the black-hole illusion and was even trying to correct for it right up to the moment he hit the trees on final approach to the runway!

Therefore, to effectively conquer these visual illusions you must not only believe that they exist, but also understand that they can trick you into believing the outside world looks fine, and thus lead you to think that no apparent hazard is present—that is why they are called *illusions*. They are not the result of some personal defect or weakness; they are normal phenomena and none of us are immune from them. "Seeing is deceiving" is true for all pilots, not just inexperienced ones.

It should be apparent that the strategies used to avoid the hazards of conducting visual approaches at night discussed in Chapter 7 are also applicable to many of the illusions discussed in this chapter. Some of those are included in the following recommendations:

- Determine beforehand if a given airport is conducive to landing illusions by using all available resources—airport publications, NOTAMs, other pilots, airport operators—to ascertain the presence of irregular-shaped runways, sloping runways or terrain, or runways conducive to black-hole conditions at night.

- Make sure your seat is adjusted so your eyes are positioned at the correct DEP. As described in Chapter 2, most commercial transport aircraft are equipped with fixed eye level indicators that can be used to adjust the captain's and FO's seats for the proper position. Contact the manufacturer if this information is not readily available in the AFM, or follow the TC *AIM* suggestions summarized in Chapter 2.

- To ensure safe obstacle clearance while conducting an approach—especially at night or in poor visual conditions—use information from your aircraft's altitude and glide path instruments (ILS, GPS, DME distance readouts, altimeter) and from the airport's visual approach slope indicator systems (e.g., VASI, PAPI).

- If only DME is available, fly a 3-degree, 4-degree or 5-degree glide path by maintaining 300, 400 or 500 feet above the runway threshold, respectively, for each NM your aircraft is away from it. For slow-flying aircraft, a 300, 400 or 500 feet-per-minute descent rate at a 60-knot ground speed yields an approximate 3-degree, 4-degree or 5-degree glide slope, respectively—just make sure you are flying at an approved AFM-recommended approach speed.

- To avoid ducking under the glide path on a precision instrument approach, continue to supplement your outside visual references with ILS glide slope information down to the runway threshold crossing height (usually 50 feet).

- Avoid long straight-in approaches (day or night), especially if you are unable to supplement outside natural visual cues with suitable altitude and glide path guidance information.

- If you suspect the possibility of an illusion, consider overflying the airport before conducting an approach to better familiarize yourself with it and the surrounding terrain.

In addition, if you are a low-time pilot, gaining experience during the day with a qualified instructor at airports with a variety of runway shapes and sizes can help if you later encounter these runways at night. The disproportionate rear-end accident rate for smaller cars when they were first introduced in North America eventually dropped off over time because drivers got used to varying-sized automobiles. The same learning occurs when you gain experience with different types of runways. Familiarity with sloped runways, sloped terrain, and runways with different dimensions creates a broader range of runway approach patterns available to you in your perceptual memory databank.

If you get caught in flat light or glassy water conditions, it's crucial to use your flight instruments to accurately confirm your aircraft's attitude and altitude until adequate outside visual references are clearly discernible. Conducting approaches in these conditions requires specialized training and extreme caution since you may be unable to acquire the visual references needed to maintain adequate terrain clearance.

To reduce the effect of a possible height illusion when flying in heavy rain, use other sources of glide path and altitude information to maintain acceptable terrain clearance—not just during the approach-to-landing phase of flight but also during the climb and en route portions. Finally, for those rare occasions when you might experience flicker vertigo, change propeller RPM or turn off the offending strobe lights to remedy the problem.

Visual illusions have deceived pilots at all levels of experience since the dawn of flight, and they will deceive you too if you fail to learn all you can about them and take the steps necessary to effectively deal with them. You can't completely trust your senses—sometimes they lie. Recognizing your limitations is half the battle in overcoming these illusions; knowing how to correct for them is the other half.

Helpful Resources

Spatial Disorientation: Visual Illusions (Part 2) is a 16-minute video produced by the Airman Education Programs branch at the FAA's Civil Aerospace Medical Institute (CAMI) that focuses on visual illusions that can lead to spatial disorientation. (www.FAA.gov/about/office_org/headquarters_offices/avs/offices/aam/cami/library/online_libraries/aerospace_medicine/aircrew/physiologyvideos/english/)

FSF ALAR Briefing Note 5.3—Visual Illusions is one of more than 30 *Approach-and-Landing Accident Reduction (ALAR) Briefing Notes* made available online to the public from the Flight Safety Foundation.

Dale Wilson has written several articles on visual illusions and night flying. Two of them are recommended below:
- *Darkness Increases Risks of Flight*, published in the November/December 1999 issue of the Flight Safety Foundation's *Human Factors & Aviation Medicine*, is an article documenting the increased risks of night flying with an explanation of some of the illusions discussed in this chapter. (FlightSafety.org/archives-and-resources/publications/human-factors-aviation-medicine/human-factors-1999)
- *Avoiding Black Holes,* published in the Fall 2004 edition of *Pilot Getaways,* explains the black-hole phenomenon. (PilotGetaways.com/backissues)

Flying in Flat Light and White Out Conditions, produced by the FAA, is available in HTML format at the *FAASTeam Learning Center Library*. (www.FAASafety.gov/gslac/alc/libview_normal.aspx?id=6844). A 27-minute FAA "Back to Basics" video by the same name is also available from the Helicopter Association International's online library. (Rotor.com/Default.aspx?TabID=239)

Notes

1. J. Robert Dille, "Visual Illusions," *Aviation, Space, and Environmental Medicine* 58 (1987): 822.

2. National Transportation Safety Board, *Safety Recommendation*, A-02-33 through -35 (Washington, DC: NTSB, October 7, 2002).

3. D. Regan, K. Beverly and M. Cyander, "The Visual Perception of Motion in Depth," *Scientific American* 241 (July 1979): 136–151.

4. Ray E. Eberts and Allen G. MacMillan, "Misperception of Small Cars," *Trends in Ergonomics/Human Factors II*, eds. Ray E. Eberts and Cindelyn G. Eberts (New York: Elsevier, 1985): 33–39.

5. Henry W. Mertens and Mark F. Lewis, "Effect of Different Runway Sizes on Pilot Performance During Simulated Night Landing Approaches," *Aviation, Space, and Environmental Medicine* 53 (1982): 463–471.

6. Ibid.

7. Helen Ross, "Mist, Murk and Visual Perception," *New Scientist* 66 (June 19, 1975): 658–660.

8. Conrad L. Kraft, "A Psychophysical Contribution to Air Safety: Simulator Studies of Visual Illusions in Night Visual Approaches," *Psychology: From Research to Practice*, eds. H.A. Pick, H.W. Leibowitz, J.E. Singer, A. Steinschneider and H.W. Stevenson (New York: Plenum Press, 1978): 363–385.

9. Martin F.J. Schwirzke and C. Thomas Bennett, "A Re-Analysis of the Causes of Boeing 727 'Black Hole Landing' Crashes," *Proceedings of the Sixth International Symposium on Aviation Psychology*, ed. R.S. Jensen (Columbus, OH: The Ohio State University, 1991): 572–576.

10. Stanley N. Roscoe, "When Day Is Done and Shadows Fall, We Miss the Airport Most of All," *Human Factors* 21 (1979): 721–731.

11. Kraft, "Psychophysical Contribution to Air Safety," 363–385.

12. National Transportation Safety Board, *Safety Recommendation*, A-02-08 (Washington, DC: NTSB, April 15, 2002).

13. Office of the NASA Aviation Safety Reporting System, "A Black Hole," *Callback* 161 (October 1992): 1.

14. P.T. Mahon, *Royal Commission to Inquire Into and Report Upon the Crash on Mount Erebus, Antarctica, of a DC-10 Aircraft Operated by Air New Zealand Limited* (Wellington, New Zealand: P.D. Hasselberg, Government Printer, 1981).

15. NTSB, *Safety Recommendation*, A-02-33 through -35.

Which Way Is Up?

SPATIAL DISORIENTATION

According to the ATC radar tapes, the airplane descended to 2,200 feet while in a right turn, then climbed to 2,600 feet and entered a left turn. It then entered another descending right turn and the airspeed, descent rate and turn rate rapidly increased. Descending at more than 4,700 feet per minute, with the airspeed exceeding redline (V_{NE}), the Piper PA-32R struck the water seven miles off the coast of Martha's Vineyard in Massachusetts, killing all three of its occupants—John F. Kennedy Jr., his wife, Carolyn, and her sister, Lauren Bessette. The weather was VFR on that warm July evening, but it was dark, it was hazy, and the pilot—Kennedy—elected to fly his Saratoga on a 30-mile direct route over open water, making it difficult for him to see a horizon that had imperceptibly blended in with the ocean. Though conspiracy theorists speculate otherwise, the NTSB determined the probable cause of this tragic accident was "the pilot's failure to maintain control of the airplane during a descent over water at night, which was a result of spatial disorientation" (NTSB Identification No: NYC99MA178).

Spatial disorientation (SD) is an incorrect orientation with respect to the position of your aircraft in space. Specifically, it is an inaccurate perception of the aircraft's attitude, direction or motion as a result of faulty or inadequate information provided by your senses—or as the FAA Advisory Circular (AC) 60-4, *Pilot's Spatial Disorientation*, puts it, it is "the inability to tell which way is up." This phenomenon usually occurs in the absence of adequate outside visual references while flying in the dark, in poor visibility or in cloud. Without sufficient input from our visual system, our brain relies more heavily on our vestibular organs and postural sensations to discern our orientation and motion. Unfortunately, when relied upon alone, these sensations can lead to a false perception of the aircraft's position in space.

Though inexperienced pilots are particularly vulnerable—Kennedy had slightly more than 300 hours—this phenomenon can afflict experienced pilots as well. An airline transport captain and commercial FO fell victim to SD while flying in IMC, causing their Learjet to impact terrain after departing Ithaca Tompkins Regional Airport in Ithaca, New York. Both pilots died in the crash. According to the NTSB, the crew failed to maintain a proper climb rate, which was the result of SD (NTSB Identification No: NYC01FA214). During a flight in IMC at night, the experienced crew of an Air Transport International scheduled air cargo flight completely lost control of a DC-8 due to SD while conducting a missed approach at the Toledo Express Airport in Swanton, Ohio. All four people aboard were killed. The captain had more than 16,000 hours, the FO more than 5,000 hours and the FE almost 22,000 hours (Report No: NTSB-AAR-92-05). Finally, SD was a major cause of Russia's third-worst aircraft accident to date. All 145 people aboard the Tupolev Tu-154 died when an experienced flight crew, conducting an approach at night in IMC, allowed the passenger jet to enter a stall and flat spin near Irkutsk, Russia.[1]

Though SD is not usually a major cause of U.S. and Canadian airline accidents, it's still a problem elsewhere. During the first decade of this century alone, almost 1,000 people lost their lives in airline accidents involving SD.[2] Data gathered by Boeing indicate that in-flight **loss of control** (LOC)—often caused by SD—is now the leading category of fatal worldwide accidents involving Western-built commercial turbojet aircraft weighing more than 60,000 pounds MCTOW.[3] Military operations are also not immune. For example, SD accounted for 11 percent of U.S. Air Force accidents over a recent 15-year period, with 69 percent of these involving fatalities.[4]

Even though SD accidents still occur in commercial flight operations, it has taken its greatest toll in the GA sector, where it was responsible for 2,355 deaths over a 17-year period (1976–1992) in the United States. More than 70 percent of those accidents occurred in IMC, half of them occurred at night and slightly more than half of them involved pilots with fewer than 500 hours' flight experience.[5] The AOPA Air Safety Institute also found that attempted VFR flight into IMC is the number one cause of fatal GA spatial disorientation accidents in the United States.[6] Though the accident rate has gradually improved—probably as a result of better education and technology—SD is still responsible for 5 to 10 percent of all GA accidents in the United States.[7]

What makes SD accidents particularly threatening is that they are almost always fatal. In fact, according to FAA statistics, more than 90 percent of GA spatial disorientation accidents result in fatalities.[8] That's because, as seen in Chapter 4, pilots who experience SD and lack the necessary instrument flying skills to safely pilot their aircraft without outside visual references either lose control of their aircraft and fly uncontrolled into terrain, or experience in-flight breakup by exceeding the design stress limits of their aircraft while attempting to recover from an unusual attitude. The study mentioned in Chapter 4 conducted by the Institute of Aviation at the University of Illinois in the 1950s confirmed this. Twenty VFR pilots, who represented a wide range of ages and flying experience but who lacked any simulated or actual instrument flight experience, flew into simulated IMC in a Beechcraft Bonanza with a qualified instructor on board. While one pilot lasted eight minutes before losing control, all of them eventually allowed the airplane they were flying to enter a dangerous flight attitude—most often a spiral dive—within an average of 178 seconds.[9]

A more recent Australian study concluded the probability of a pilot experiencing at least one episode of SD during his or her flying career was 90 to 100 percent. You've probably heard the old adage that goes, "There are two types of pilots—those who've landed with the gear up, and those who will." The ATSB study concluded, "There are those who've experienced SD and those who will."[10] Whether you are an experienced commercial pilot or just beginning to spread your wings, if that conclusion is correct, it's crucial that you understand the nature of SD, how to avoid it and, just as importantly, how to effectively overcome it should you experience it.

What Is Spatial Disorientation Anyway?

The three primary systems that provide the information you need to correctly perceive movement and orientation in space are the visual, vestibular and somatosensory systems. When you are spatially disoriented you may experience illusions corresponding to one or more of these systems in the form of visual illusions, **vestibular illusions** or **somatosensory illusions**. Visual illusions were introduced in Chapter 7 on night flying and were also discussed in Chapter 8. In this chapter we explore the major vestibular and somatosensory illusions resulting from SD.

The study of SD began more than a century ago when the Austrian physicist Ernst Mach discovered that self-motion could be perceived not only visually but also through the vestibular organs located in the inner ear. The vestibular apparatus, located in the bony labyrinth of the inner ear, consists of three semicircular fluid-filled canals that respond to angular acceleration, and the otolith bodies that respond to linear acceleration. The semicircular canals are arranged approximately at right angles to each other and closely correspond to the three flight axes of pitch, yaw and roll. A slight bulge in each canal contains receptor hairs that are attached to the cupula that occupies the diameter of the canal. Since the fluid has inertia and resists movement, the cupula flows through the fluid with acceleration of the head or body (i.e., canals) in any of the three axes. The bending of the hairs in the cupula triggers the receptor cells, which in turn communicate to the brain that acceleration in a particular plane is occurring.

The same principle of inertia applies to the otolith organs of the saccule and utricle; however, a different structure is at work. A dense, flat, plate-like membrane overlies an array of tiny sensory hairs that, when deflected by head tilt or forward acceleration or deceleration, activate the sense receptors connected to them.

It's important to understand that the vestibular apparatus can only detect angular and linear *acceleration*. Once the acceleration stops, the fluid in the semicircular canals catches up to the body, and the tiny hairs are no longer deflected. Likewise, the membranes in the otolith bodies also stop their movement once acceleration stops. As a result, we sense acceleration and deceleration, but do not detect steady-state motion.

The third major way in which you perceive orientation in space is through the somatosensory system. Sometimes referred to as the proprioceptive or kinesthetic sense, when using it for orientation in flight it is commonly known by pilots as "flying by the seat of the pants." Like the vestibular system, the somatosensory receptor cells of the skin, muscles, joints and tendons respond only to accelerations. Gravity is the major acceleration detected, but in flight these postural sensations are unable to distinguish between gravity and angular or linear accelerations (or G-forces). For example, when the acceleration detectors in the semicircular canals equilibrate in a prolonged level turn, only the somatosensory sense is detected. The centrifugal force in a level turn combines with gravity to produce one resultant force; depending on how steep the bank, you will perceive that you are now in a wings-level straight climb.

The vestibular and somatosensory acceleration detectors are often quite accurate for normal earth-bound activities that usually involve a series of accelerations and decelerations. Also, while maintaining a constant velocity in an automobile or aircraft, our visual system is dominant and overrides the limited inputs provided to the brain from the other two systems, giving us an accurate perception of our speed and movement. However, when flying in poor visibility, at night or in cloud, there is insufficient information in the external environment (such as the natural horizon or terrain features) for our visual system to provide an accurate perception of our orientation in space. It is in these situations that our vestibular system and seat-of-the-pants sensations can create SD in the form of false perceptions (illusions).

Though many nonvisual illusions resulting from SD have been experienced by pilots, only the major ones will be discussed here. Some illusions involve both systems simultaneously, while most involve either the semicircular canals or the otolith bodies of the vestibular system. Some illusions—mostly those involving the otolith organs—also involve sensations provided by the somatosensory system.

The Leans

One of the most common vestibular illusions is known as **the leans**. Since the hairs in the cupula are unable to sense rolling motion below the threshold of about 2 degrees per second, an inadvertent rolling movement below this threshold will go unnoticed. When you eventually detect this motion through input provided by the flight instruments, you will likely correct for the resultant bank angle by rolling the wings level at a roll rate that is above the detection threshold. This will now deflect the cupula in the semicircular canals, giving a strong sensation of entering a turn in the *opposite* direction. Recognizing a conflict between these body sensations and the instrument indications, you might instinctively try to resolve it by *leaning* your body in the direction of the original bank.

Dale Wilson, one of the authors of this book, clearly remembers his first experience with this illusion. It happened on his first instrument flight test, which was also the first time he ever flew in actual IMC. While tuning the radio and listening to ATIS, his Piper Warrior entered a slow roll (below the 2-degree-per-second threshold). After redirecting his gaze to the flight instruments he realized he had entered a left turn but was dumbfounded as to why he hadn't "felt" this. After rolling the airplane back to a wings-level attitude (above the 2-degree roll threshold) he felt a strong sensation of turning to the right. Even though his flight instruments indicated he was flying straight and level, his body wanted to lean back to the left to reconcile these conflicting sensations. This actually happened two more times during the flight, and in case you're wondering, he did manage to overcome it and successfully pass his checkride.

Graveyard Spin and Spiral

Since the semicircular canals only react to accelerations and not steady velocities, false sensations can be experienced during prolonged angular motions such as turns, spins or spirals. After about 10 to 20 seconds of constant angular motion, the fluid in the semicircular canals catches up to the speed of the canal walls and the cupula returns to its neutral resting state, giving the sensation of zero motion or acceleration.

Two potentially deadly illusions, as their names imply, are the **graveyard spin** and **graveyard spiral**. These somatogyral illusions (*soma*=body; *gyral*=circle) can cause you to make inappropriate control movements based on false information. For example, if for whatever reason you find yourself in a prolonged spin without the benefit of outside visual references to determine your aircraft's attitude (at night or in IMC), after about 20 seconds you will stop perceiving the spinning sensation. The fluid in the semicircular canals has caught up to the speed of the canal walls, the sensors in the cupula have returned to their resting state, and no motion is perceived. When you attempt to recover from the spin you will experience a strong sensation of spinning in the opposite direction, even though the flight instruments indicate otherwise. If you respond to these erroneous feelings and ignore the instruments, you may attempt to recover from this falsely perceived spin and inadvertently re-enter a spin in the original direction.

A similar situation can occur when equilibrium is attained during a prolonged spiral dive. When you level the wings to recover from the spiral you will experience a strong sensation of entering a turn in the opposite direction. If you make control corrections based on these sensations instead of the information provided to you by your flight instruments, you could re-enter the original spiral dive. This illusion is likely to be fatal if adequate outside visual references are not made available in time.

This is what happened to the crew of a Beechcraft King Air B200 while flying in IMC on a winter evening in Colorado. Carrying members of the Oklahoma State University (OSU) basketball team to Stillwater, Oklahoma, the aircraft experienced an electrical failure shortly after reaching cruising altitude. This rendered the pilot's side flight instruments unusable except for the airspeed and turn-and-slip indicator. Not long after the failure, the B200 entered a descending turn to the right that was, according to the NTSB, "consistent with a graveyard spiral resulting from pilot spatial disorientation." The horizontal stabilizer was aerodynamically overloaded due to the pilot's pull-up maneuver, which led to an in-flight breakup. All 10 people on board were killed (Report No: NTSB/AAR-03/01).

Coriolis Illusion

Another potentially deadly phenomenon is the **Coriolis illusion**. After the semicircular canals have equilibrated during a prolonged turn, you might tilt your head to look for a chart or pick up a pencil off the floor. This brings a second canal into the axis of rotation, causing its cupula to deflect, and moves the original canal out of the axis of rotation, causing its cupula to deflect in the opposite direction as its canal fluid begins to slow down. Because of the conflicting sensations of acceleration in one axis and deceleration in the other, this cross-coupling illusion produces strong sensations of motion in the third. So powerful is this illusion that pilots who've experienced it have reported strong pitch and bank illusions, sensations of tumbling forward, and even nausea.

We mentioned in Chapter 6 that the FAA provides physiology training to civilian pilots in Oklahoma City. One component of this training is familiarization with spatial disorientation, including a practical demonstration of the Coriolis illusion in a Barany chair (similar to a swivel office chair with arm rests). A volunteer is securely strapped into the chair and asked to put her head down and close her eyes while the instructor turns the chair at a constant, relatively slow rate for at least 20 seconds. When the instructor stops the chair, the volunteer is asked to lift up her head and open her eyes. The effect on participants is disquieting: Their entire body moves as they try to stabilize themselves to a visual scene that is erratically rotating about them, and they feel as if they will topple off the chair. Some have even felt nauseous for several hours afterward.

Somatogravic Illusions

The false climb illusion was introduced in the chapter on night flying as a hazard during dark-night takeoffs in otherwise clear weather conditions. Officially named the **somato-gravic illusion** (*soma* = body; *gravic* = gravity) by aviation physiologists, it arises from misinterpreted inputs from both the otolith bodies of the vestibular apparatus and the sensations provided by the somatosensory receptors when the aircraft is subject to rapid acceleration when flying in poor visibility, in cloud or in the dark. As indicated earlier, movement of the plate-like membrane in the otoliths provides the same sensations whether your head is tilted up or is accelerated in a straight line during a takeoff or a go-around. This illusion is further strengthened by your somatosensory receptor system since it cannot distinguish between gravitational and linear accelerations. These seat-of-the-pants sensations combine gravitational and linear accelerations into one resultant force, causing you to incorrectly interpret the result of the backward force of inertia and gravity as the gravitational vertical (Figure 9-1). If you are deceived by this illusion, you will correct for these false sensations by inadvertently placing the aircraft in a nose-down attitude, causing your aircraft to fly right into the ground!

Unfortunately, pilots are still being fooled by this. According to the Canadian TSB, the captain of a Fairchild Metroliner III operating on a scheduled commercial flight into Northwest Regional Airport, in Terrace, British Columbia, experienced this illusion while conducting a missed approach in IMC. In spite of verbal call-outs from the FO that the aircraft was descending, the captain continued the descent and the aircraft crashed into the trees, killing all seven occupants (TSB Report No. 89H0007). This was also the experience of a Gulf Air Airbus A320 that recently crashed into the Persian Gulf on a dark August night after the captain experienced the false climb illusion during a go-around at Bahrain International Airport. All eight crew members and 135 passengers on board died in the crash.[11] Finally, it appears that the captain of an Armavia Airlines Airbus A320 that recently crashed during a missed approach at night in IMC at Sochi International Airport in Russia was also the victim of the somatogravic illusion. He initiated a go-around on short final and climbed above 1,600 feet, then pushed the nose down, causing the jet to descend at 4,300 fpm; the airplane crashed into the Black Sea killing all 113 people on board.[12]

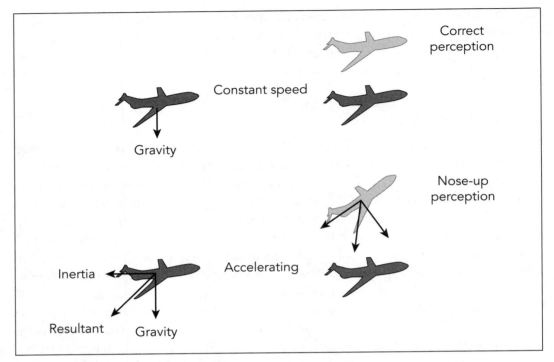

Correct
perception

Constant speed

Gravity

Nose-up
perception

Inertia

Accelerating

Resultant   Gravity

Figure 9-1. False climb (somatogravic) illusion.

Incidentally, the somatogravic illusion also works in reverse, as the crew of a Lockheed C-5 Galaxy discovered when the aircraft inadvertently entered a stall after the pilots probably corrected for a perceived nose-down attitude brought about from a rapid deceleration in level flight. Fortunately, they were able to safely recover.[13]

Other variations of the somatogravic illusion involve significant changes in aircraft load factor, or G-force. Primarily applicable to fighter and aerobatic pilots whose maneuvers regularly involve departure from normal +1G flight, these illusions can trick any pilot—especially one struggling to recover from an inadvertent unusual attitude resulting from SD.

Elevator Illusion

Positive and negative accelerations are measured along three axes: G_X (longitudinal axis), G_Y (lateral axis) and G_Z (vertical axis). We usually experience $+G_Z$ (pronounced "positive gees") acceleration in the vertical axis (headward acceleration, or *eyeballs down*) mostly from the increased load factor involved in turns or from a rapid updraft; occasionally we experience $-G_Z$ acceleration (footward acceleration, or *eyeballs up*) that would occur during a rapid downdraft.

Positive vertical G_Z increase the strength of the gravitational acceleration and provide more pull on the otolith organs, which only serves to intensify any somatogravic-type illusion. The **elevator illusion** is one such example: in the absence of outside visual references, a strong updraft increases the strength of the gravitational acceleration, creating the sensation of a nose-high attitude. If you correct for this by pitching the nose down while flying close to the ground, a CFIT accident is possible. Of course, a nose-down pitch illusion is possible during a sudden downdraft.

Inversion Illusion

Another variation of the somatogravic illusion occurs during an abrupt level-off from a climb that involves pulling $-G_Z$ prior to the level-off. Usually restricted to aerial combat or acrobatic flight maneuvers, this situation creates a strong nose-up pitch illusion as the aircraft accelerates and returns to $+G_Z$ flight—so strong, in fact, that an illusion of flipping backward to an inverted position can result. Unfortunately, as is the case with the standard somatogravic illusion, you may attempt to overcome this **inversion illusion** by pushing the airplane into a dive.

G-Excess Illusion

False sensations of pitch or bank can also result when head movement is initiated in high $+G_Z$ flight. Like other somatogravic illusions, the increased gravitational acceleration and greater pull on the otolith organs lead to this **G-excess illusion**, which, if not recognized, will cause you to make inappropriate control responses. A Mooney M20K that crashed 19 seconds after takeoff at dusk in IMC was observed to level off over Runway 34 at Greater Binghamton Airport in New York, accelerate, and then climb up into an indefinite ceiling in fog. The NTSB determined the cause of this fatal accident was SD, likely involving a G-induced illusion (NTSB Identification No: NYC08FA039).

Vertigo

Sometimes pilots use the term "**vertigo**" to describe SD, but its definition is actually much narrower in scope and, as a type of SD, refers to a spinning or twirling sensation either of yourself or your surroundings. One type of severe vertigo involves involuntary back-and-forth eye movements known as **nystagmus**. A person who suddenly stops after rotating on a piano stool for several seconds experiences this. The same phenomenon can happen in flight during recovery from prolonged angular maneuvers involving high rates of rotation; involuntary nystagmus causes your visual scene to appear to move back and forth. A person who experiences the Coriolis illusion will also suffer from the disorienting effects of nystagmus. Obviously, this phenomenon would make recovery from an unusual attitude even more difficult.

Classifying Spatial Disorientation

Aviation physiologists have identified three different types of SD. If you experience the first type—**Type I (unrecognized) SD**—you will be unaware of your disorientation and will base control of your aircraft on a false perception of your attitude. The somatogravic (false climb) illusion is an example of this, and an accident is likely if you fail to recognize it in time.

With **Type II (recognized) SD** you are aware of your disorientation, and at first may suspect there is something wrong with your flight instruments. You will likely experience disorientation stress as you fight your vestibular and/or somatosensory sensations in an attempt to trust and fly by the instruments—a process aviation physiologists call "establishing visual dominance."

Unfortunately, if you experience **Type III (incapacitating) SD**, even though you are aware of your disorientation, you might not be able to properly take positive control of your aircraft.[14] Extreme aircraft attitudes, complicated by disorienting visual symptoms such as nystagmus, can make it difficult, if not impossible, to recover from this type of SD.

Overcoming Spatial Disorientation

So much for the bad news. The good news is that these illusions can be completely overcome if you know how to pilot your aircraft solely by reference to the flight instruments. Pilots who have survived a serious disorientation episode have often reported that when they broke out of IMC into visual conditions, their SD *immediately* disappeared and they were able to correctly orient themselves using natural outside visual references. Therefore, you can avoid SD—or successfully defeat it if you experience it—by learning to rely on cockpit instrumentation designed to replicate the natural references of the outside world.

Trust Your Instruments

When properly scanned and interpreted, your aircraft's flight instruments give you an accurate depiction of the aircraft's orientation in space. Real-time information about altitude, airspeed, heading, rate of climb/descent, and rate of turn is provided to you by these instruments. Especially helpful is the attitude indicator that substitutes the natural horizon with an artificial one, providing you with a *direct* indication of the aircraft's pitch and bank attitude. Pilots who are proficient at using the flight instruments to fly their aircraft—and there are thousands of them who do so every day—have learned to trust what the instruments are telling them and to ignore vestibular and somatosensory sensations.

Of course, to trust the flight instruments they must be trustworthy. It's important, therefore, to check them for proper operation and accuracy *before* you commit to the air. Not doing so could be disastrous, as the captain of a Cessna Citation II discovered shortly after takeoff from Mountain View Airport in Missouri. In a hurry not to miss his IFR departure clearance time, he rushed his departure procedures—the time from engine start-up to takeoff was only about two minutes—and didn't allow adequate time for the gyros, which power the attitude and horizontal situation indicators, to spool up. The airplane crashed in IMC shortly after takeoff, killing all three people on board (NTSB Identification No: DCA83AA005).

Just because you've confirmed that the flight instruments are operating correctly before you launch doesn't mean they won't fail once you're airborne. An instrument failure can be catastrophic. That's what happened to the Beechcraft King Air B200 carrying members of OSU's basketball team (the accident mentioned earlier); a complete captain's side instrument failure occurred. However, the airspeed and turn-and-slip indicator, along with four of the copilot's side flight instruments, continued to function properly. An instrument failure also occurred in India's second-worst accident to date. Flight 855, a Boeing 747, crashed after takeoff killing all 213 people on board after the captain's attitude director indicator failed, causing confusion and subsequent SD. The captain of the Air India flight failed to use the correct instruments for control and rolled the aircraft over into a 108-degree left bank.[15]

Though flight instrument failure is a rare occurrence, every pilot must be prepared for it; that's why "partial panel" practice is required in every instrument training syllabus and for every instrument proficiency check. In the event of an instrument failure, it's important to know how to cross-check, recognize, isolate and ignore the failed flight instrument (or instrument cluster) and be proficient at flying with the remaining backup instruments. You should also be familiar with your aircraft's approved *Emergency and Abnormal Procedures Checklist* before you have to pull it out and use it.

Maintain Instrument Flying Proficiency

Maintaining proficiency in instrument flying techniques is a must if you plan to fly IFR. If you know an IFR flight is approaching, you can stay sharp by spending some time in the simulator or a flight training device (FTD), on a basic or advanced aviation training device (ATD), or even flying your own computer-based flight simulator. If it's been a while since you've received instruction in unusual attitude recovery using only the flight instruments, you should refresh those skills as well. Also, if your aircraft is equipped with a **glass cockpit**, make sure you have received proper training and are proficient using backup flight instrumentation if the electronic systems fail; a study by the NTSB in 2010 found that GA aircraft equipped with digital glass cockpits had a higher fatal accident rate than those with conventional steam gauge flight instrumentation.[16]

When flying in IMC it's important to avoid large abrupt head movements, especially during prolonged turning maneuvers where the semicircular canals have equilibrated. It goes without saying that you should also avoid extreme pitch and bank attitudes; this will prevent tumbling not only of the instrument gyros but also your inner-ear gyros. Avoiding excessive or unusual attitudes also reduces the possibility of SD symptoms that could arise when pulling vertical G_Z to recover from them.

When flying VFR in poor visibility or at night, supplement outside references with cockpit flight instruments to avoid not only vestibular and somatosensory illusions but also visual illusions (discussed in Chapter 8). You should continue to rely on the cockpit flight instruments until the outside references are distinguishable enough to maintain accurate orientation. You should also avoid flying at night if you are not proficient in instrument flying techniques or are not night-current.

Avoid VFR Flight into IMC

A commercial pilot, who also held a CFI certificate, planned to depart for a business meeting in Klamath Falls, Oregon. But when he and his passenger arrived at the Seaside Municipal Airport on the morning of the meeting, it was blanketed with fog. The rest of the route was good VFR, and the pilot probably believed he could safely climb through the fog and make it to the meeting. Unfortunately, in the two minutes it took to climb through the fog, the pilot likely suffered from SD before entering a graveyard spiral and crashing into a house near the airport. Not only were the two people on board killed, but so were three children who were in the house at the time of the crash (NTSB Identification No: LAX08FA256).

As you read in the introduction to this chapter, the number one cause of fatal SD accidents is attempting VFR flight into IMC, and those who attempt to fly in such conditions without the instrument flying skills required to do so have, on average, 178 seconds to live. Therefore, if you are a non-instrument-rated VFR pilot, or have no recent instrument experience, your only option is to completely avoid flying into IMC altogether. Sometimes this is not as easy as it sounds. In Chapter 4 we alluded to the fact that it is difficult for pilots, especially less experienced ones, to accurately determine actual ceiling and visibility values from the cockpit of a moving aircraft. This is a crucial skill if you are going to avoid flight into IMC. We also mentioned that you should establish your own personal weather minimums to guide you when the weather goes bad. However, establishing these minimums is of no value unless you resist pressure—both from yourself and from others—to continue flight when you encounter conditions below those minimums. In case you inadvertently find yourself in IMC, you should know how to conduct a 180-degree standard-rate turn using only the flight instruments. If equipped, you should also know how to make an autopilot-assisted 180. Review Chapter 4 on VFR flight into IMC for more suggestions.

Obtain an Instrument Rating

Finally, if you are a non-instrument-rated pilot, why not gain some added insurance and versatility from your aircraft by obtaining an instrument rating? AOPA's statistics indicate that only 17 percent of accidents resulting from VFR flight into IMC involve instrument-rated pilots.[17] Not only are you more likely to use your IFR ticket in marginal weather, but if you inadvertently find yourself in the soup, you will have the skills necessary to get out of it and avoid becoming an accident statistic.

Helpful Resources

The FAA's Civil Aerospace Medical Institute (CAMI) provides several free videos that graphically illustrate various aspects of spatial disorientation. (www.FAA.gov/about/office_org/headquarters_offices/avs/offices/aam/cami/library/online_libraries/aerospace_medicine/aircrew/physiologyvideos/english/) and (www.FAA.gov/about/office_org/headquarters_offices/avs/offices/aam/cami/library/online_libraries/aerospace_medicine/sd/videos/)

- *Spatial Disorientation: Vestibular Illusions (Part 1)*
- *G-Excess Effect*
- *Grave Yard Spiral*
- *Inversion Illusion*
- *Otolith Illusions*
- *Pitch-Up Illusion*
- *Semi-Circular Illusion*

The AOPA Air Safety Institute provides several free online interactive courses on IFR flying and two in particular that can help you better understand and manage the spatial disorientation threat. (www.AOPA.org/education.aspx)

- *Single-Pilot IFR*
- *Real Pilot Stories: Vacuum Failure in IMC*

Spatial Disorientation is an AOPA Air Safety Institute *Safety Advisor* that can assist you in managing the threat of spatial disorientation. (www.AOPA.org/asf/publications/advisors.html)

Aeromedical Factors, Chapter 16 in the FAA's latest *Pilot's Handbook of Aeronautical Knowledge* (FAA-H-8083-25), provides an excellent overview of spatial disorientation. (www.FAA.gov/regulations_policies/handbooks_manuals/aviation/pilot_handbook/)

Notes

1. Flight Safety Foundation, Aviation Safety Network, *Accident Description* (Vladivostokavia Air, Tupolev-154, Burdakovka, Russia, July 4, 2001). Available online at Aviation-Safety.net/database/record.php?id=20010704-0.

2. David Learmount, "Pilot Disorientation Accidents Have Become a Phenomenon," *Flightglobal* (January 29, 2010). Available online at www.Flightglobal.com/articles/2010/01/29/337743/pilot-disorientation-accidents-have-become-a-phenomenon.html.

3. Boeing Commercial Airplanes, *Statistical Summary of Commercial Jet Airplane Accidents: Worldwide Operations*, 1959–2012 (Seattle, WA: Boeing, August 2013).

4. Terence J. Lyons, William Ercoline, Kevin O'Toole and Kevin Grayson, "Aircraft and Related Factors in Crashes Involving Spatial Disorientation: 15 Years of U.S. Air Force Data," *Aviation, Space, and Environmental Medicine* 77 (July 2006): 720–723.

5. W.E. Collins and C.S. Dollar, *Fatal General Aviation Accidents Involving Spatial Disorientation: 1976–1992*, DOT/FAA/AM-96/21 (Washington, DC: FAA Civil Aeromedical Institute, August 1996).

6. AOPA Air Safety Institute, *Safety Advisor—Spatial Disorientation: Confusion That Kills*, Physiology No. 1 (Frederick, MD: AOPA ASI, August 2004).

7. Federal Aviation Administration, *Spatial Disorientation: Why You Shouldn't Fly by the Seat of Your Pants*, AM-400-03/1 (Washington, DC: FAA). Available online at www.FAA.gov/pilots/safety/pilotsafetybrochures.

8. Ibid.

9. Leslie A. Bryan, Jesse W. Stonecipher, and Karl Aron, "180-Degree Turn Experiment," *Aeronautics Bulletin* 52.11 (Urbana, IL: University of Illinois, 1954).

10. David G. Newman, *An Overview of Spatial Disorientation as a Factor in Aviation Accidents and Incidents*, Report B2007/0063 (Canberra City, Australia: Australian Transport Safety Bureau, 2007). Available online at www.ATSB.gov.au/publications/2007/b20070063.aspx.

11. Bahrain Civil Aviation Affairs, *Aircraft Accident Investigation Report, Gulf Air Flight GF-072, Airbus A320-212, REG. A40-EK August 23, 2000, at Bahrain* (Kingdom of Bahrain: Bahrain Civil Aviation Affairs). Available online at www.BEA.aero/docspa/2000/a40-ek000823a/htm/a40-ek000823a.html.

12. M. Lacagnina, "Into the Black Sea: A Go-Around Goes Awry in Sochi, Russia," *AeroSafety World* (October 2007): 44–49. Available online at FlightSafety.org/aerosafety-world-magazine/past-issues.

13. Fred H. Previs and William R. Ercoline, "Spatial Disorientation in Aviation: Historical Background, Concepts, and Terminology," *Spatial Disorientation in Aviation*, eds. Fred H. Previs and William R. Ercoline (Reston, VA: American Institute of Aeronautics & Astronautics, 2004): 1–36.

14. U.S. Department of the Army, *Aeromedical Training for Flight Personnel*, Training Circular No. 3-04.93/FM 3-04.301 (Washington, DC: U.S. Dept. of the Army, August 31, 2009): 9.1–9.2.

15. Flight Safety Foundation Aviation Safety Network, *Accident Description* (Air India, B-747, Arabian Sea, off Bandra, January 1, 1978). Available online at Aviation-Safety.net/database/record.php?id=19780101-1.

16. National Transportation *Safety Board, Safety Study: Introduction of Glass Cockpit Avionics Into Light Aircraft*, NTSB/SS-01/10 (Washington, DC: NTSB, March 9, 2010). Available online at www.NTSB.gov/safety/safetystudies/SS1001.html.

17. AOPA ASI, *Spatial Disorientation*.

Where Am I? 10

The pilots of American Airlines Flight 965 must have been puzzled. Only a minute before, the captain had programmed the FMS computer to fly their Boeing 757 direct to the Rozo nondirectional beacon (NDB)—the final approach fix for the VOR/DME instrument approach to Runway 19 at Cali, Colombia. It was located straight ahead, but both crew members were surprised when they discovered that the airplane had turned to a heading about 90 degrees to the left of their desired course. The FO turned the aircraft back to the correct heading, but both pilots were unaware that it had drifted significantly off course and was now heading toward mountainous terrain. The onboard GPWS alerted the pilots of the rising terrain ahead, and they responded by adding full thrust and pitching the airplane into a steep climb attitude. Unfortunately it was too late; the airplane crashed into the side of Mount El Deluvio at 8,900 feet, killing all but four passengers of the 164 people on board.[1]

The crash of Flight 965 that December evening—the first fatal accident involving a B-757 and Colombia's worst aviation disaster to date—is an example of a controlled flight into terrain (CFIT) accident. You've seen several examples of these types of accidents throughout this book, but this final chapter is devoted exclusively to this deadly threat. A CFIT accident involves a properly functioning aircraft under the control of a fully qualified pilot or flight crew that is unintentionally flown into terrain (or water or obstacles). The pilots are either unaware of the impending collision or, as in the case of Flight 965, if they do become aware, it's too late to avoid it.

CFIT has historically been the number one cause of aviation fatalities in commercial airline accidents. A study conducted in the 1990s found that it had claimed the lives of more than 9,000 passengers and airline crew members since commercial passenger jet operations began in the mid-1950s.[2] CFIT was still the leading cause of worldwide airline fatalities between 1987 and 2005—responsible for the loss of 3,735 lives[3]—but since then the number of fatalities has been slowly decreasing, in large part due to better education and the use of improved CFIT-avoidance technology. In-flight loss of control (LOC) has recently taken over as the leading category of worldwide fatal airline accidents, but over a recent 10-year period (2001 through 2010) CFIT was still responsible for the loss of 1,007 lives in commercial airline operations.[4]

All of the above statistics involve Western-built commercial turbojet aircraft with MCTOWs greater than 60,000 pounds; they do not include accidents involving small commuter, air taxi and GA aircraft, which compose the majority of the world's aircraft fleet. In fact, 71 percent of CFIT accidents involve smaller aircraft with fewer than 10 passenger seats.[5] Researchers at the Volpe National Transportation Systems Center conducted a study of GA CFIT accidents that occurred over a 12-year period in the United States. During that period, 1,260 Part 91 CFIT accidents occurred—an average of 105 per year—resulting in 1,789 fatalities.[6] The study also found that even though CFIT was responsible for less than 5 percent of all GA accidents, it was to blame for more than 17 percent of all GA accident fatalities. Because of the higher speeds often involved at impact, CFIT accidents have a higher proportion of fatalities, similar to VFR flight into IMC, which is a major cause of these types of accidents.

This high fatality rate is also seen in commuter and air taxi operations. For example, only 17 percent of 351 commuter and air taxi accidents that occurred in Alaska over an eight-year period (1991–1998) were the result of CFIT, but it was responsible for 59 percent of all commuter and air taxi fatalities and 55 percent of pilot deaths.[7] In fact, the study found that a passenger traveling on a commuter or air taxi flight in that state had 12 times the risk of dying in a CFIT accident than a non-CFIT accident. Another alarming study found that 71 percent of all fatal aircraft accidents between 1990 and 1998 in Alaska were the result of CFIT.[8]

Civilian aviation is not the only sector threatened by this hazard: military operations have also been plagued by it. It is estimated that approximately 20 percent of military accidents and about 40 percent of military aviation fatalities are the result of CFIT. For example, over a 10-year period, CFIT was responsible for the loss of 202 U.S. Air Force and Navy/Marine Corps aircraft, 431 lives, and more than $3.1 billion.[9,10]

CFIT Risk Factors

One would think CFIT is a problem mostly for flights that routinely fly close to terrain or water, such as helicopter operations, low-level military flights, aerial applications (crop dusting) or aerial surveillance. For example, as this chapter was being written, a U.S. Coast Guard Sikorsky MH-60 Jayhawk crashed into the waters off the coast of Washington State, killing three of its four crew members. The helicopter hit power cables that were strung between James Island and the coastal community of La Push at an approximate altitude of 114 feet MSL. While without question these types of flights operate at higher levels of risk, the statistics above indicate that no segment of aviation is immune from the threat of CFIT—pilots and aircraft of all types are vulnerable. A variety of factors increase the probability of experiencing CFIT, and a number of these were present in the crash of American Airlines Flight 965. The following discussion highlights some of these factors.

Atmospheric Conditions

The majority of CFIT accidents occur in conditions that make it difficult or impossible to see terrain: cloud, fog, poor visibility, or nighttime. Although scattered clouds were reported near Cali, the weather at the time of the accident was essentially clear—but it was a dark and moonless night, making it next to impossible for the crew to see the rising terrain ahead. Research scientists at the Netherlands National Aerospace Laboratory (NLR) studied 156 worldwide fatal commercial (major, regional and air taxi) airplane CFIT accidents that occurred during a seven-year period. They discovered that 87 percent of the accidents occurred in IMC and almost half took place at night.[11] In its Advisory Circular 61-134, *General Aviation Controlled Flight into Terrain Awareness*, the FAA reports that more than half of all GA CFIT accidents occur in IMC. It's difficult to avoid terrain that you can't see.

Geographic Location

The probability of experiencing a CFIT accident varies depending on the part of the world in which you're flying. American Airlines Flight 965 crashed in a country deemed to be a high-risk area. The NLR study found that compared to North America, flying over the southern continent increases the risk of experiencing a CFIT accident by a factor of five. Africa has the highest CFIT accident rate, increasing the risk eightfold. In some areas the probability is estimated to be as much as 80 times greater![12]

The facilities and level of service that pilots are used to in North America are not always available elsewhere. Two important support systems we take for granted are terminal approach radar and minimum safe altitude warning systems (MSAW). The latter automatically alerts controllers when an aircraft has descended below the minimum safe altitude. Flight 965 didn't have the benefit of these. Had Cali airport been equipped with either, the controller might have been able to warn the crew of the airplane's course deviation and its dangerously close proximity to the terrain. According to the NLR report, no approach radar was available for 37 percent of the fatal commercial airplane approach and landing CFIT accidents, and in 25 percent of the cases no approach lights or approach slope indicating systems (e.g., VASI, PAPI) were available.

Even if available, approach slope indicators, approach and runway lighting systems, and navigation aids in other parts of the world may not be in compliance with ICAO standards. There's also no guarantee that instrument approach procedures conform to the exacting standards of the United States (*U.S. Terminal Instrument Procedures*), Canada (*Criteria for the Development of Instrument Procedures*) or ICAO (*Procedures for Air Navigation Services—Aircraft Operations*). An improperly designed instrument approach procedure was partly to blame for the CFIT accident involving a U.S. Air Force CT-43 (Boeing 737) on an NDB approach to Dubrovnik Airport in Croatia. All 35 people on board died in the crash, including U.S. Secretary of Commerce Ron Brown.[13]

Finally, although the international language of aviation is English, the ability of the world's air traffic controllers to understand and speak it varies significantly from region to region. The official accident report reveals that the captain of Flight 965 and the Cali controller were operating on different assumptions, partly because the two didn't share the same primary language. The controller later testified that he thought they had both understood each other's communications, but a couple of the captain's requests didn't make sense to him. He said if the pilots spoke Spanish he would have questioned them, but because of his limited skills in non-aviation English he was unable to do so. Even though Colombian investigators concluded the controller's actions didn't contribute to the crash, had he and the crew of Flight 965 shared a common primary language, this accident might have been prevented.

Phase of Flight

Though CFIT accidents can happen during any phase of flight, the majority occur during the approach to landing. Flight 965 crashed 25 miles northeast of the airport while conducting a straight-in instrument approach for Runway 19 at Cali. According to the NLR study, most of the sample CFIT accidents—almost 70 percent—occurred during the approach and landing phase of flight (from the top of descent to landing).

Type of Instrument Approach

Conducting a **nonprecision approach**—especially at night or in IMC—increases the probability of CFIT. Almost 60 percent of the fatal commercial airplane approach and landing

CFIT accidents in the NLR study involved nonprecision instrument approaches. A precision approach, such as an ILS approach, provides *precise* vertical guidance to fly at a stabilized approach angle; a nonprecision approach, such as a VOR, VOR/DME, localizer/back course (LOC/BC) or NDB, does not. Especially troublesome are approaches with multiple step-down fixes; pilots can easily become confused and descend below the applicable altitude before reaching the fix. According to the Flight Safety Foundation, it is five times safer to fly a precision instrument approach than a nonprecision one.[14] The pilots of Flight 965 were conducting a VOR/DME approach into Cali. They were originally cleared for the ILS approach for Runway 01, but were given the option of the VOR/DME approach for Runway 19; because they were running late, they took the straight-in nonprecision approach.

Unexpected Changes

Unplanned changes during flight have contributed to accidents. Such alterations have caught pilots off-guard, leaving them ill-prepared to adapt to the change. Such was the case for the pilots of Flight 965. They were originally set up for an ILS approach for Runway 01, which would have involved flying south past the airport and conducting a procedure turn for a landing to the north. However, the flight departed Miami almost two hours late and the crew was worried the flight attendants might not obtain a sufficient rest period, which would have disqualified them for duty the next day. So when the controller offered them a straight-in approach for Runway 19, they welcomed it. However, since they were not expecting such a clearance, they knew they needed to expedite their descent if they were going to successfully conduct the VOR/DME approach for Runway 19. The pilots now found themselves rushed, and whenever a job is rushed, mistakes are bound to happen. A critical task that prepares both crew members for the approach is an **approach briefing**, yet there is no indication on the CVR that either crew member conducted one. The accident report also indicates that no approach checklist was completed, nor were the pilots communicating with each other as effectively as they should have been.

Communication and Automation

The flight had originally been cleared via the "Rozo One Arrival," which would take it 29 NM south-southwest of the Tulua VOR in a valley keeping the B-757 away from the high mountains on either side, to the Rozo NDB located 2.6 NM north of the airport. From there the pilots were to fly 10 NM south to the Cali VOR, then conduct a procedure turn on the ILS approach and land on Runway 01 from the south. However, when the controller cleared them to Cali, the captain thought he meant *direct* Cali, and since the controller never corrected the read-back (the controller replied "Affirmative" to the pilot's read-back instead) the captain commanded the FMS to take them *direct* to Cali. Unfortunately, when he did this, the FMS logic erased the "Rozo One Arrival" information from the control display unit (CDU). After receiving an amended clearance for the VOR/DME approach for Runway 19, the captain requested a clearance direct to Rozo to get course guidance to the runway. The controller then cleared the flight to fly the "Rozo One Arrival," but because of the ambiguous nature of the communication between them, the captain thought he was cleared *direct* to the Rozo NDB.

Unfortunately, at this point the FMS required more careful attention than the captain was able to give it. When he called up the "R" NDB (Rozo) on the FMS, the CDU listed several to choose from, in order of closest proximity to the aircraft. So the captain selected and executed the first "R," which commanded the FMS to fly the aircraft to the new waypoint. What he didn't know, however, was there were actually two "R" NDBs in Colombia, each with the exact same identifier, and the one he selected was the Romeo ("R") NDB located approximately 150 NM northeast (to the left) of their position. Apparently the Rozo NDB never even appeared on the CDU screen; for it to do so it would have had to be entered using its full name of "Rozo" instead of "R." The American Airline's SOP specified that the data must be verified and approved by the other crew member before a change in the FMS was executed; unfortunately, there was no evidence that this took place. Had the captain followed the SOP, perhaps the critical mistake that caused their aircraft to turn off course might not have occurred.

The pilots appeared to over-rely on the autopilot and FMS to fly and navigate their airplane while they were managing all the details associated with the change of runway; they didn't realize it had turned left until more than a minute had passed. The last-minute change of runways dramatically increased the crew's workload, and this—along with the communication misunderstandings between the captain and controller, inadequate communication between the two flight crew members, and misuse and misunderstanding of the FMS—all conspired to cause Flight 965 to stray off course toward higher terrain.

Steering Clear of CFIT

In the early 1990s, the Flight Safety Foundation led an international task force made up of experts from more than 30 different industry and government organizations to develop strategies to combat the CFIT threat. Within three years they produced the *CFIT Education and Training Aid* (see Helpful Resources at the end of this chapter). Although a number of factors can lead to this type of accident, the task force concluded the fundamental cause of most CFIT accidents is lack of **situational awareness** on the part of flight crews. Various definitions of situational awareness exist, but at its most basic level it means having sufficient *awareness* of all the necessary aspects of your *situation* to conduct a safe flight. The task force specifically cited the lack of two types of situational awareness that can lead to a CFIT accident: *vertical* positional awareness and *horizontal* positional awareness (since if you don't know where you are horizontally, you won't know your vertical proximity to terrain).

The question then becomes, "What can I do to maintain my positional awareness and avoid a CFIT accident?" It turns out there are many proven strategies to use. The rest of this chapter highlights several behaviors you can engage in to beat this hazard, but the strategy that will help you the most can be summarized as *don't completely trust yourself.* As illustrated in accident examples throughout this book, pilots don't always have an accurate mental picture of their situation. When it comes to avoiding CFIT, it's important that you double-check *everything*—information received from your flight and navigation instruments, from your charts, from ATC, and even from the pilot flying next to you. A healthy dose of skepticism never hurt anybody in aviation. Remember what we mentioned earlier: just because you're not paranoid, doesn't mean they're not out to get you.

Be Prepared for Your Flight

As is the case for many threats in aviation, avoiding a CFIT accident begins on the ground. During your preflight preparations, pay particular attention to minimum terrain and obstruction clearance altitudes for your route of flight. As discussed in Chapter 7, if you are flying VFR, use the terrain and obstacle information provided on VFR sectional charts. If flying IFR, you should familiarize yourself with all applicable minimum IFR altitudes including: minimum en route IFR altitudes (MEA) and minimum obstruction clearance altitudes (MOCA), minimum off-route altitudes (MORA) or off-route obstruction clearance altitudes (OROCA), minimum safe altitudes (MSA) and, for your anticipated instrument approaches, minimum descent altitudes/heights (MDA/H) or decision altitudes/heights (DA/H). You should also study and thoroughly understand the instrument approach procedures for your destination and alternates, especially at unfamiliar airports.

Avoid VFR Flight into IMC

Many CFIT accidents are the result of VFR pilots attempting flight into IMC. It was noted in Chapter 4 that VFR flight into IMC is a major killer in aviation. For example, most of the fatal commercial aviation accidents in Alaska were the result of CFIT, and most of those were due to pilots continuing VFR flight into IMC.[15] The NLR study also found that almost 20 percent of 156 worldwide fatal commercial CFIT accidents were the result of inadvertent flight from VMC into IMC. Chapter 4 explores the problem of VFR into IMC in depth and offers tips on how to avoid it.

Be Aware of Your Altitude

You should familiarize yourself with applicable minimum safe altitudes before your flight, and should also know what they are at all times during flight. The NLR study found that published minimum safe altitudes were not maintained in 35 percent of CFIT accidents. Though hard to believe, misread altimeters have also led to CFIT accidents. For example, pilots can mistake an indication of 6,000 feet on the altimeter for 16,000 feet; in fact, this is what investigators speculated might have happened to the crew of a United Airlines B-727 that descended into Lake Michigan. Although the probable cause was never conclusively determined, investigators hypothesized that the crew may have erroneously misread "6,000" feet indicated on the altimeter for "16,000" feet.[16]

A mid-1960s study found that pilots took up to three times longer to interpret the traditional three-pointer altimeter, and were up to 11 times more likely to misread it compared to other types of altimeters.[17] There are still many three-pointer altimeters out there, as well as newer ones that are less likely to be misinterpreted. But no matter the altimeter improvements, interpretation errors by pilots still occur, so be careful.

Many CFIT accidents might also be avoided if pilots use correct altimeter settings. A Bell 206 helicopter emergency medical services (HEMS) flight, operating VFR on a dark night in VMC, collided with upsloping terrain while cruising in level flight. Examination of the wreckage indicated the altimeter was incorrectly set, resulting in a reading that was 310 feet higher than the aircraft's actual altitude (NTSB Identification No: CHI04FA107). While conducting a nighttime nonprecision approach into a small airport on the coast of British Columbia, the crew members flew a Learjet 35 into the ocean 8 NM short of their destination. All five people aboard perished. The TSB concluded that the pilots had incorrectly set their altimeters to 30.17 in Hg, a full inch above the required setting of 29.17, and as a result their aircraft was consistently 1,000 feet below the required altitudes for the approach (TSB Report No: A95P0004).

For most aircraft, the altimeter is the only instrument that indicates the aircraft's altitude; therefore, in IMC or at night, it's crucial that you get it right. Make a habit of reading back all altimeter settings and double-check to make sure the correct value is always set in the altimeter. It's the responsibility of each crew member in a crewed cockpit to confirm the correct setting.

Be aware that even with the correct setting in the subscale, in nonstandard atmospheric conditions your altimeter will lie to you. Especially critical is an OAT that is much colder than standard temperature: Your aircraft's actual altitude (or *true* altitude) will be lower than what the altimeter indicates. To ensure adequate obstruction clearance while flying in IMC or at night in these conditions, you should consult the "Cold Temperature Error Table" in the FAA *AIM* (or the "Altitude Correction Chart" in the TC *AIM*) and add the appropriate altitude correction value to the published minimum IFR approach altitudes (e.g., DME arc, procedure turn, final approach fix [FAF], MDA/H or DA/H).

Finally, make sure you know what units of measurement are used when flying outside of the United States and Canada. Many countries report altitude in meters, wind speed in meters per second, and altimeter settings in millibars (mb), or hectopascals (hPa). After receiving an altimeter setting of "998" from Brussels approach control, the crew of a B-767 set 29.98 in Hg in the altimeter subscale; the only problem was that the controller meant 998 mb (hPa)! This resulted in the airplane descending at least 350 feet below the minimum safe altitude indicated on the approach chart (ASRS Report No: 565879). Be aware that not all countries use a transition altitude of 18,000 feet (FL180) for changing from flight levels (standard pressure of 29.92 in Hg) to altitudes based on local altimeter settings. Also, even though altimeter settings for most countries provide altitude above MSL (station pressure converted to MSL, sometimes called "QNH"), some jurisdictions actually use station pressure settings that provide altitude above field elevation (called "QFE"). Make sure you know which is which.

Always Know Your Location

After *aviate,* a pilot's next responsibility is to *navigate.* We read earlier that good CRM involves the use of all available resources to achieve safe and efficient flight, and nowhere does this apply more than to the task of navigating from one location to the other. When flying VFR, you must know how to identify natural and human-made ground features with those on the chart (pilotage), but you should also be able to use your planned headings, courses and distances to predict where you should be at pre-calculated times (dead reckoning). This latter skill is especially important when flying in remote areas with limited terrain features and when flying at night. These skills apply to IFR flights as well, except when in IMC you must use DME distances, cross-bearings and other radio-navigation sources to verify your position.

Whether you are flying VFR or IFR, always try to use more than one source of information to confirm your position, but beware of the **confirmation bias**. This is the tendency to look only for information that confirms what you believe rather than information that may prove you wrong. For example, one of this book's authors remembers a cross-country flight with one of his students. He set his heading over an oval racetrack even though it was located 10 miles from the one he intended to use. Instead of looking for features on the chart that didn't match the terrain he was flying over, or suspecting his significantly premature arrival at the checkpoint (he didn't use dead reckoning), he latched on to the first ground reference he saw to confirm where he thought he was. It's a better strategy to try and *disconfirm* your position, and if you're unable to, then you're probably located where you think you are.

If you have the opportunity, consider upgrading your aircraft with GPS-based navigation equipment. Terrain is difficult to avoid if you can't see it; however, many of these systems have the ability to display terrain and obstacles on a moving map. Since a picture is worth a thousand words, this technology can increase your positional awareness—especially in the dark or when flying in the soup. One caution: Since these systems usually provide flight instrumentation in addition to navigation information, make sure your backup systems and plans are in place should the entire system fail (e.g., electrical failure). Dark displays can make aviating and navigating a difficult challenge.

Avoid Finger Trouble

Just as an incorrect value can be set in the altimeter, finger trouble can affect other important systems as well. The crew members of a twin-engine corporate turbojet found this out shortly after takeoff when they experienced a windshield crack and engine vibration. They requested a clearance back to the airport, and while the captain was flying and managing the engine problem, the FO programmed the FMS and dialed in the ILS frequency for Runway 26. After the FO called "localizer alive" and reported to ATC that they were established on the approach, they began their descent only to receive a GPWS warning and a call from ATC to immediately climb to avoid terrain. It turns out they were not east of the airport as they should have been, but 12 miles south of it. The FO had inadvertently programmed the FMS for Runway 08 (ASRS Report No: 843379).

Throughout this book we've stressed the importance of neither completely nor blindly trusting yourself—or your copilot—when it comes to setting critical values into your navigation radios and other systems. You need to double-check all data entries and settings, and if flying in a crewed flight deck, have your fellow crew member verify their accuracy. By the way, you should be especially vigilant when changing communications frequencies while on a vector toward higher terrain at night or in cloud; make sure you tune in the correct frequency, and if you fail to get a response in a timely manner, go back to the previous frequency. If necessary, initiate a lost communications procedure and climb to the minimum safe altitude.

Communicate, Communicate, and Communicate

We mentioned earlier that communication difficulties exist in countries where English isn't the official language; however, even in English-speaking countries, barriers to effective communication occur daily in the cockpit. Most of these do not result in accidents, but occasionally they do. In Chapter 7 we introduced the tragedy of Eastern Air Lines Flight 401, a Lockheed L-1011 that flew under controlled fight into the terrain of the Florida Everglades northwest of Miami International Airport. What we didn't mention was the Miami approach controller knew something the crew members didn't: Less than a minute before the accident, his radar screen indicated their airplane was at 900 feet, which was 1,100 feet below their assigned altitude of 2,000 feet. In response, the controller asked, "Eastern, ah, four oh one, how are things comin' along out there?" but didn't specifically mention the low altitude (Report No: NTSB-AAR-73-14). Had the controller pointed this out—he believed it was a false altitude reading, since those were common back then—perhaps the crew's attention would have been directed to the aircraft's low altitude.

A search of the ASRS database reveals the majority of reports have to do with communication issues between pilots and controllers: Pilots have misunderstood clearances, have incorrectly read them back (the **read-back problem**) or have missed them altogether; controllers have failed to correct inaccurate read-backs (the **hear-back problem**), have issued incorrect clearances, and have even issued unsafe clearances.

When under the guidance of air traffic controllers, don't assume they will always be able to steer you clear of terrain. A student and instructor were conducting an instrument approach at Bremerton National Airport in Washington State and were cleared to turn to "290 degrees and maintain 2,000" on the missed approach. Had the error not been caught by the instructor, flying that heading would have put them on a collision course with terrain (ASRS Report No: 528109). In another incident, an air carrier crew was told to maintain 7,000 feet until reaching a certain intersection in Nevada, but that put the Boeing jet 2,000 feet below the minimum safe IFR altitude; the crew was cleared to a higher altitude too late and had to alter heading toward lower terrain to ensure safe terrain clearance (ASRS Report No: 555274). If you cannot comply with a clearance, or if you feel it is unsafe, it's your responsibility to question ATC and, if necessary, reject the clearance altogether.

Don't assume what the pilots of TWA Flight 514 did. When they were cleared for the VOR/DME approach for Runway 12, about 44 miles from the Washington Dulles International Airport, without any altitude restrictions, they *assumed* their Boeing 727 was under radar surveillance and once "cleared for the approach" they could descend to the FAF altitude of 1,800 feet as depicted on the chart. This was a deadly assumption: when they reached that altitude prior to the fix, they crashed into the west slope of Mount Weather in Virginia, killing all 92 souls onboard (Report No: NTSB-AAR-75-16).

Responsibility for terrain clearance does not belong to ATC, it belongs to you—the pilot-in-command. Even after experiencing hundreds of successful flights under the capable hands of air traffic controllers, you need to avoid the tendency to become complacent by over-relying on their ability to keep you safe. Controllers make mistakes, too, so don't blindly trust them.

Ineffective **extra-cockpit communication** between pilots and controllers has contributed to CFIT accidents; poor **intra-cockpit communication** between crew members has as well. A United Airlines DC-8 ran out of fuel and crashed in a residential neighborhood near Portland, Oregon, because the FO and FE failed to communicate the seriousness of the jet's fuel status to the captain (Report No: NTSB-AAR-79-7). As we saw in Chapter 1, the world's worst accident to date occurred in part because the FO and FE failed to effectively communicate their doubts to the captain about a takeoff clearance. Better communication between crew members might also have prevented the crash of Flight 965 in Colombia: According to the CVR, the FO neglected to conduct an approach briefing, the captain failed to ask the FO to verify the accuracy of information selected in the FMS and, as the PM, he also failed to make the required verbal call-outs on the approach.

Effective verbal communication is a pillar of good CRM. Without it, a crew will not perform as an effective team. In the busy, noisy and sometimes stressful environment of the cockpit, successful communication can be a challenge. Some messages can have more than one meaning (e.g., "to three zero" versus "two three zero"), letters can sound the same (e.g., B, C, D, E, G, P, T, V), nonstandard phraseology is sometimes used (e.g., "seven five zero zero" for altitude versus "seven thousand five-hundred") and radio transmissions are sometimes garbled or stepped on by others' transmissions. Good communication does not come naturally for many, but it is a skill that can be learned with practice.

To make sure you and your listeners (ATC, other flight and cabin crew members, etc.) are on the same page, you need to practice good communication skills. This involves not only following all the standard recommended procedures such as using standard phraseology, proper annunciation and pronunciation, the phonetic alphabet and your correct call sign, but also practicing other effective communication strategies such as active listening, information seeking, asking questions, restating or clarifying the message, and conveying information in a forthright and sometimes assertive manner. If frustration or interpersonal conflict results, it's important to focus on *what* is right, not *who* is right. A study published in the early 1980s found that flight crews who communicated less tended not to perform as well as crews who communicated more.[18] A flight deck where frequent, open and effective communication is expected is crucial to a successful and uneventful flight.

Follow Standard Procedures

Chapter 3 referenced FAA Advisory Circular 120-71, *Standard Operating Procedures for Flight Deck Crewmembers*, in the context of the important role the PM plays in watching for any aircraft performance degradation from airframe ice accumulation. Those who fly for a living know how important SOPs are for safe flight operations; those who don't should. SOPs are established procedures that, when followed, ensure the highest possible level of safety. At a minimum, they contain the collective wisdom of an organization's most experienced pilots, but often they consist of industry-wide best practices on how virtually every aspect of a flight should be conducted to ensure safe operations. Some examples of what SOPs prescribe include: specific tasks to accomplish for each phase of flight, from preflight to after-landing taxi; what to do in abnormal or emergency situations; how to use checklists and automation; the duties of the PF and PM in a crewed cockpit, including the use of briefings and standard call-outs; and procedures to follow for maintaining altitude awareness.

AC 120-71 notes that almost half of all commercial CFIT accidents involve the failure of flight crews to comply with published SOPs, and that sometimes there is an altogether lack of adequate SOPs available. If you fly single-pilot operations and want to reduce the level of risk in your flying, contact a flight instructor or the FAA (or TC in Canada) and either will be glad to assist you in developing your own personal SOPs.

Be Extra Vigilant When Conducting the Approach

Nowhere are the benefits of following SOPs more important than during the approach and landing. By now you've seen how this part of the flight poses the greatest risk—it accounts for more than a third of the accidents but less than four percent of the flight time. That's because of all the phases of flight, the performance standards required in conducting a successful approach and landing are at their highest and, because the margin of error is so low, any substandard performance increases the odds of an accident. The following discussion draws upon the guidance provided in FAA AC 120-71, *Standard Operating Procedures for Flight Deck Crewmembers*, describing instrument approaches and other procedures primarily applicable to two-person crews flying transport category aircraft. However, even if you don't fly in a crewed cockpit, or aren't an IFR-rated pilot, you can still benefit from the wisdom of maintaining a sterile cockpit, conducting approach briefings, using checklists, making critical call-outs, and flying stabilized approaches.

Maintain a Sterile Cockpit

Several times in this book we have stressed the importance of complying with the sterile cockpit rule—even if you don't fly for a living as a commercial airline pilot. As you recall, this rule prohibits engaging in nonessential activities—including conversations not pertinent to the task at hand—that could distract you from completing the essential duties required for the safe operation of your aircraft during the critical phases of flight. There are no more critical phases of flight than the approach and landing. Therefore, since they require your highest levels of attention to accomplish safely, you should do all you can to avoid any distractions—self-induced or otherwise—that would hinder that task. Olympic athletes talk about how they need to stay "in the moment" to achieve their high levels of performance; pilots need to as well. Yes, you must be able to think ahead and be prepared for contingencies such as a go-around or missed approach, but flying a successful approach—especially in challenging conditions—demands all your resources, your skills, and the ability to stay focused. Some airline pilots say out loud, "I, thou, here and now" when they are having difficulty staying in the moment. With only yourself and one other in a crewed cockpit while conducting an approach, neither of you should be thinking about what *just* happened or too much about what *will* happen, but should focus on what is happening right *now*. This obviously can't be done if you are not maintaining a sterile cockpit.

Review and Brief the Approach

To adequately prepare for the most challenging phase of flight, it is standard airline practice for pilots to "brief the approach" before commencing it, ideally before the top-of-descent. The PF should conduct the briefing by reviewing and verbally explaining how he or she plans to fly the procedure. At a minimum, the briefing should include destination airport weather, altimeter setting (and any temperature corrections needed), where the descent will begin, inbound and outbound courses, important fixes, minimum safe altitudes on the approach (e.g., procedure turn, intermediate approach fix, FAF, MDA/H, DA/H), descent rates, target airspeeds, and the missed approach procedure. Conducting an approach briefing contributes to a major goal of CRM: working together as a team. It ensures both pilots are on the same page (if something doesn't sound right, the PM tells the PF), allows the PF to direct greater attention to the task of actually flying the approach and provides a standard by which the PM can monitor the accuracy of, and assess deviations from, the approach.

Fly Stabilized Approaches

The majority of approach-and-landing accidents are the result of unstabilized approaches.[19] An unstabilized approach (too high, too low, too fast or too slow) significantly adds to your workload, decreasing your odds of a successful landing. Combined with other required tasks—such as configuring the aircraft, completing checklists and communicating with ATC, to name a few—flying an unstabilized approach demands more attention than you may be able to give and could lead to a critical or fatal error.

According to the advisory circular (FAA AC 120-71), a **stabilized approach** involves flying at a relatively constant approach angle and rate of descent (normally no greater than 1,000 feet per minute) down to the flare-out point within the touchdown zone of the runway. An accurate track and glide path angle should be flown from the glide path intercept point, or by the FAF inbound, with no more than *normal* bracketing corrections needed to maintain them both. Unless otherwise specified in the SOP, the approach should be stabilized by no lower than 1,000 feet height above touchdown (HAT) in IMC or 500 feet HAT in VMC, the airplane should be in the landing configuration (gear and flap), airspeed and power changes should be within the normal range as per your aircraft operating manual, and all checklists and most call-outs should be completed. You should execute a missed approach if these parameters cannot be met. Making it a habit to stabilize the approach earlier, rather than later, gives you the ability to focus on the task at hand and increases the likelihood of a successful landing.

Use Checklists and Call-Outs

Several air carrier accidents have occurred because pilots have forgotten to complete a checklist (forgetting to set the flaps for takeoff, for example). Along with read-and-do **checklists**, airline pilots conduct memorized **flow checks** during all pertinent phases of flight. However, these flow checks are followed up with the appropriate checklist, which adds an extra level of safety by ensuring nothing has been forgotten. Of course they're useless if you forget to use them in the first place! In addition, they're only effective if you actually visually (or physically) check and confirm that the checklist action has been accomplished. You can also get into serious trouble when a checklist is interrupted. When this happens, it's important to return to where you left off and finish it. The FAA advisory circular (AC 120-71) also recommends that checklists be accomplished no lower than 1,000 feet HAT in IMC or 500 feet HAT in VMC.

Finally, whether you're monitoring your own performance or that of your fellow pilot, making important **call-outs** during the approach also can prevent a CFIT accident. The NLR study found that failure to complete checklists and call-outs were causal factors in many CFIT accidents.[20] Completing these is an essential aspect of effective CRM and is codified in most airline SOPs, but they can be beneficial for single-pilot operations as well. Numerous call-outs are made by both the PF and PM during flight (critical altitudes, exceeding navigation tolerances, autopilot mode changes, etc.), but those relating to altitude awareness on the approach are fairly standard. The PM usually makes the following call-outs: when the localizer and glide path are alive and then when they have been captured; crossing altitudes when passing approach fixes; specific feet-to-go reminders above HAT, MDA/H, or DA/H (e.g., "one thousand to go," "one hundred to go," etc.); when at minimums and whether visual contact is established or not when the aircraft reaches those minimums. At the FAF inbound the PM usually cross-checks the accuracy of both sets of flight instruments including altimeters and radar altimeter altitude.

Chapter 1 introduced the Eastern Air Lines Flight 212 accident, a DC-9 that collided short of Charlotte/Douglas International Airport while the crew was conducting an approach. This accident resulted in the creation of the sterile cockpit rule in the United States in 1981. Unfortunately, not only did the crew members engage in extraneous conversations, they flew an unstabilized approach, didn't cross-check their instruments, and failed to make the SOP-required call-outs at the FAF. Whether flying by yourself or as part of a team, making such call-outs will increase your positional awareness and help you to avoid premature contact with the ground.

Avoid Nonprecision Approaches

We mentioned that the CFIT risk increases fivefold when flying a nonprecision approach compared to one that provides precise vertical guidance (e.g., ILS). That's partly because the chance of busting an altitude is greater when you have to keep track of altitudes associated with step-down fixes. If you have a choice between flying this type of approach and following a glide path down to the touchdown zone of the runway, why not choose the latter? We also discussed the positional awareness benefits of flying an aircraft with GPS-based navigation equipment; another benefit of some of these systems is the ability to conduct precision-like approaches. The increasing use of wide area augmentation system (WAAS) area navigation (RNAV) instrument approaches in the United States, Canada and elsewhere allows pilots of properly certified GPS-equipped aircraft to conduct ILS-like instrument approaches. These vertical navigation (VNAV) approaches provide a constant angle of descent, enabling a more stabilized approach as well as more accurate vertical guidance, thereby reducing the chance of experiencing a CFIT accident.

Avoid Visual Approaches at Night

As you saw in Chapter 7, conducting visual approaches at night (or in low visibility) also increases your chance of experiencing a CFIT accident. The crew and passengers of a Boeing 737 were saved by the GPWS on a clear dark night when the aircraft began descending toward a 4,682-foot mountain peak after the crew accepted a visual approach to Runway 11L at Tucson International Airport (ASRS Report No: 541269). If you are conducting an instrument approach at night—even in clear weather—and you are given the option of a visual approach, don't take it. But if you do, continue to use the guidance provided by the instrument approach procedure to avoid untimely contact with terrain.

Manage the Automation

The captain of American Airlines Flight 965 selected the wrong waypoint in the FMS, and both he and the FO failed to adequately monitor the autopilot as it turned the Boeing 757 in the wrong direction. Failure to effectively manage automated systems has contributed to several tragic CFIT accidents. One of these involved an Air Inter Airbus A320 in France. The crew of Flight 148 was conducting a VOR/DME Runway 05 approach into Strasbourg International Airport when the A320 collided into the side of Mt. Sainte-Odile, killing 87 of the 96 people on board. It appears the crew may have experienced **mode confusion** by setting the autopilot to descend in the wrong mode. Instead of selecting "-3.3" in the *FPA* (flight path angle) mode in the flight control unit (FCU), which would have given them a safe 3.3 degree glide path (about 800 fpm), "33" was selected in the *V/S* (vertical speed) mode, yielding a descent rate (3,300 fpm) four times greater than what was required.[21,22]

The advent of modern computerized glass cockpits with features designed to reduce pilot workload and improve situational awareness has also brought unintended consequences of increased complexity and the increased likelihood of confusion on the part of pilots. For example, automatic flight control systems can operate in several "pitch" modes (e.g., vertical speed mode, airspeed mode, pitch angle mode). To safely operate within prescribed parameters, the system will automatically change modes on its own; however, sometimes it does so with minimal or no notification to the pilot. Even older autoflight systems can produce unintended consequences.

One of the authors of this book was flying a King Air into a busy international airport and was given descents in multiple steps. The autopilot was on and was just capturing one altitude when a new altitude was given by ATC. The new altitude was selected in the altitude window, but because the autopilot had silently shifted from vertical speed mode to altitude capture mode, this produced a sudden and unanticipated steep nose-down attitude before the pilot was able to disconnect the autopilot and hand-fly the airplane back to a normal descent profile. Remember, it is your responsibility to thoroughly understand how the automated systems in your aircraft operate and how the various modes function.

Both the American Airlines B-757 and Air Inter A320 CFIT accidents demonstrate another side effect of automation: the phenomenon known as **automation complacency**. It appears the pilots in both accidents over-relied on the automation to safely fly their airplanes. You've undoubtedly heard the expression that flying is "99 percent sheer boredom and 1 percent sheer terror." The boredom often comes because the autopilot does what you want it to do 99 percent of the time, so with every flight, your level of trust increases and you begin to relax your monitoring role—especially when busy with other tasks. The pilots of both of these flights were very busy preparing for the approach, but it appears they lost their positional awareness as they allowed the autopilot to fly their airplane. At various points throughout this book we've advised you to not completely trust yourself, the pilot sitting next to you, or ATC—and now, the automated systems in your aircraft should be added to that list. Your priorities are always to aviate, navigate and communicate—in that order. Delegating your most important task—flying the airplane—to the autopilot without adequately monitoring it could kill you.

Manage the Unexpected, Don't Let It Manage You

Expect the unexpected in flight. Interruptions, distractions and changes are bound to occur. Sometimes they're intentional (e.g., choosing a more favorable runway); sometimes they're not (e.g., diversion to another airport due to weather). Either way, it's important not to let them keep you from your primary job of flying the airplane and maintaining safe terrain clearance. The crew of American Airlines Flight 965 was given the choice of runways and the crew of Air Inter Flight 148 was not, but according to the accident investigation reports, in both instances a last-minute unexpected runway change significantly increased their workload, causing them to rush their preparations for the approach. Unfortunately, because of this, the job of minding the store received less attention than it should have.

When you are dealing with a changing situation, avoid becoming so mentally over-loaded with managing the change that flying the airplane and positional awareness suffer. You must also make a realistic assessment of your ability to accomplish the tasks needed to achieve a safe outcome within the limited timeframe given you; if you don't think you can, then buy more time (e.g., slow down, request a hold, conduct a go-around, or don't accept a clearance). Like most pilots, it's difficult to admit—to others *and* to yourself—when you might be getting "behind" the airplane. When things begin to snowball, rather than trying to save face and risk getting in deeper, admit you're having trouble and consider a Plan B. Often that's the first step in avoiding an accident.

Comply with Terrain Warnings

There is another reason to fly aircraft equipped with GPS-based navigation equipment, especially if that equipment includes terrain awareness and warning systems (TAWS)—it saves lives. As mentioned in Chapter 7, TAWS provides terrain avoidance information to the pilot. Required for commercial airliners in the United States after the Flight 965 Cali accident (and now mandatory in Canada), these systems provide pilots with sufficient warning (through visual and aural alerts) of potentially hazardous terrain in time to take effective corrective action to prevent a CFIT accident.[23] It is the latest version of GPWS. Since the use of GPWS was mandated in the United States after the 1974 TWA Flight 514 CFIT accident near Dulles, it has dramatically reduced the incidence of CFIT accidents among commercial air carriers. According to some estimates, the risk was reduced by 95 percent for aircraft equipped with this first-generation terrain warning system.[24] However, it had its limitations. False alarms and nuisance warnings caused pilots to delay their response to GPWS alerts or ignore them altogether, which unfortunately sometimes led to CFIT accidents. Also, because the system relied on downward-looking radar altimeter (or *radio* altimeter) information, it could only give a few seconds of warning if the terrain elevation rose too steeply—not enough time for pilots to successfully execute an escape maneuver.

Advances in computer technology and the widespread use of GPS satellite navigation paved the way for the newer enhanced GPWS (EGPWS). This system combines aircraft GPS position with a computer-stored terrain and obstacle elevation database, providing forward-looking capability and giving flight crews significantly earlier advance warning of threatening terrain. EGPWS meets the criteria of TAWS, but "TAWS" is the more generic term that is inclusive of other systems that meet its criteria. Expensive EGPWS is not an option for most small GA aircraft, but low-cost systems meeting TAWS criteria are now available and are being used in many aircraft. At a minimum, TAWS installed in air transport airplanes provide alerts for the following situations: reduced terrain clearance, imminent terrain impact, a premature descent, an excessive rate of descent, a negative climb rate or altitude loss after takeoff, and when at 500 feet above the terrain or nearest runway elevation (voice call-out "Five hundred") during a nonprecision approach. Most systems also provide a terrain awareness moving map display that shows surrounding terrain or obstacles relative to the aircraft.[25]

As a last line of defense, TAWS is only effective if you heed its warnings. Commercial airline pilots are trained to conduct a **CFIT escape maneuver** when they hear the EGPWS "Pull up! Pull up!" or "Terrain, Terrain!" alerts. However, several incidents and accidents indicate that it's important to adhere to your company's SOP or manufacturer's recommended procedures for its use. Industry safety experts recommend that you not waste valuable time diagnosing the problem; you instead should immediately respond to TAWS (EGPWS) warnings. The FAA advisory circular (AC 120-71) recommends that upon hearing the aural "Pull up" or "Terrain" warnings, the PF manually set full firewall power (thrust), disconnect the autopilot, pitch the nose up at 3 degrees per second to a 20-degree pitch attitude (increase pitch attitude if warning continues while respecting stall buffet/stickshaker and avoiding a stall), retract speed brakes (but leave flaps and gear), and climb to a safe altitude. If you fly in a TAWS-equipped aircraft, follow your organization's SOP or the aircraft manufacturer's recommended procedures for its use.

CFIT is a major killer in aviation. Like most accidents, it's seldom caused by any one factor—usually there are several that, when combined, conspire to cause an accident. Your main strategy to combat this threat is to always maintain your positional awareness. Even if you think you know where you are—laterally or vertically—just remember: *don't believe everything you think.* To quote a phrase made popular by the late U.S. President Ronald Reagan: "Trust, but verify." Double-check everything and don't completely or blindly trust anyone, including yourself. When you're interrupted, distracted, or get preoccupied with anything that causes you to lose your positional awareness, do whatever it takes to get it back. If it doesn't feel right, don't be too proud to execute a missed approach or climb to a higher altitude to assess your situation. Remember, altitude is often your friend.

Helpful Resources

CFIT Education and Training Aid, developed by Boeing, ICAO, FAA and the Flight Safety Foundation, is an excellent resource for pilots and managers and provides best practice strategies to reduce the CFIT risk. (www.FAA.gov/training_testing/training/media/cfit/volume1/titlepg.pdf)

General Aviation Controlled Flight into Terrain Awareness is an FAA advisory circular (AC 61-134) that also includes a copy of the Flight Safety Foundation's CFIT Checklist, a useful tool for operators who want to assess their risk for CFIT. (www.FAA.gov/regulations_policies/advisory_circulars/index.cfm/go/document.information/documentID/22907)

Terrain Avoidance Plan (TAP) is an article written by Bruce Landsberg in *AOPA Online* (July 2004) that provides tools to VFR pilots for avoiding CFIT. There is also a link to its *Terrain Avoidance Plan Safety Brief.* (www.AOPA.org/asf/asfarticles/2004/sp0407quiz.html)

AOPA Air Safety Institute free online interactive courses can increase your knowledge of a variety of aspects designed to help you reduce your CFIT risk. (www.AOPA.org/asf/online_courses)

- *GPS for VFR Operations*
- *GPS for IFR Operations*
- *IFR Insights: Regulations*
- *IFR Insights: Charts*
- *Say It Right: Mastering Radio Communication*
- *Single-Pilot IFR*
- *IFR Chart Challenge: VOR Approach*
- *IFR Chart Challenge: ILS Approach*
- *IFR Chart Challenge: RNAV Approach*

NASA's Aviation Safety Reporting System (ASRS) mines incident report data supplied by pilots, controllers, and others to identify deficiencies in the aviation system and improve aviation safety. The December 2013 issue of *Callback*, their safety newsletter, uses incident reports to educate readers about the challenges of successfully managing automated systems on aircraft. You can access specific issues, as well as information on how to subscribe to this monthly electronic publication, online. (asrs.arc.NASA.gov/publications/callback.html)

Automation, Chapter 7 in the FAA's new *Risk Management Handbook* (FAA-H-8083-2), can help you better understand how to manage the automated systems in your aircraft. (www.FAA.gov/library/manuals/aviation)

Notes

1. Aeronautica Civil of The Republic of Colombia, *Aircraft Accident Report: Controlled Flight Into Terrain, American Airlines Flight 965, Boeing 757-223, N651AA, Near Cali, Colombia, December 20, 1995* trans. Peter B. Ladkin (Santafe de Bogota D.C., Colombia: Aeronautica Civil of The Republic of Colombia, September 1996). Available online at sunnyday.MIT.edu/accidents/calirep.html.

2. Federal Aviation Administration, *Controlled Flight Into Terrain Education and Training Aid* (Washington, DC: FAA, September 1997).

3. Boeing Commercial Airplanes, *Statistical Summary of Commercial Jet Airplane Accidents: Worldwide Operations, 1959–2005* (Seattle, WA: Boeing, May 2006).

4. Boeing Commercial Airplanes, *Statistical Summary of Commercial Jet Airplane Accidents: Worldwide Operations, 1959–2010* (Seattle, WA: Boeing, June 2011).

5. William B. Scott, "New Research Identifies Causes of CFIT," *Aviation Week & Space Technology* 146 (June 17, 1996): 70–71.

6. Melissa J. Bud, Peter Mengert, Stephen Ransom and Mary D., *General Aviation Accidents, 1983–1994: Identification of Factors Related to Controlled-Flight-Into-Terrain (CFIT) Accidents, Final Report*, DOT/FAA/AAR-100-97-2 (Cambridge, MA: U.S. Department of Transportation, Volpe National Transportation Systems Center, July 1997).

7. Timothy K. Thomas, Jan C. Manwaring, and George A. Conway, "Controlled Flight Into Terrain Accidents Among Commuter and Air Taxi Operators in Alaska," *Aviation, Space, and Environmental Medicine* 71.11 (November 2000): 1098–1103.

8. Larry Bailey, Linda M. Peterson, Kevin W. Williams and Richard C. Thompson, *Controlled Flight Into Terrain: A Study of Pilot Perspectives in Alaska*, DOT/FAA/AM-00/28 (Washington, DC: Federal Aviation Administration, Office of Aviation Medicine, August 2000).

9. Gary Bell, "Making 'Saves' With GPWS," *Approach* (July–August 2008): 10–12.

10. Michael L. Moroze and Michael P. Snow, *Causes and Remedies of Controlled Flight Into Terrain in Military and Civil Aviation* (Wright-Patterson AFB, OH: Human Effectiveness Directorate, Air Force Research Laboratory, 1999).

11. R. Khatwa and A.L.C. Roelen, "An Analysis of Controlled-Flight-Into-Terrain (CFIT) Accidents of Commercial Operators 1988 Through 1994," *Flight Safety Digest* 15 (April–May 1996): 1–45.

12. David Hughes, "Safety Group Highlights CFIT Risk for Regionals," *Aviation Week & Space Technology* 140 (May 9, 1994): 46–51.

13. Linda D. Kozaryn, *Air Force Releases Brown Crash Investigation Report* (Washington, DC: American Forces Press Service, June 13, 1996).

14. Edward H. Phillips, "Surge in CFIT Accidents Sparks Renewed Concerns," *Aviation Week & Space Technology* 157 (December 9, 2002): 50–51.

15. Thomas, Manwaring, and Conway, "Controlled Flight Into Terrain," 1098–1103.

16. National Transportation Safety Board, *Aircraft Accident Report: United Air Lines Inc., B-727, N7036U, in Lake Michigan, August 16, 1965*, File 1-0030 (Washington, DC: NTSB, December 19, 1967).

17. R. Chernikoff and P.N. Ziegler, *An Experimental Evaluation of Four Types of Altimeters Using Both Pilots and Enlisted Men Subjects*, NRL-6232 (Washington, DC: Naval Research Laboratory, December 18, 1964).

18. H. Clayton Foushee and Karen L. Manos, "Information Transfer Within the Cockpit: Problems in Intracockpit Communications," *Information Transfer Problems in the Aviation System*, NASA TP-1875, eds. C.E. Billings and E.S. Cheaney (Moffett Field, CA: NASA-Ames Research Center, September 1981): 63–71.

19. Flight Safety Foundation, *ALAR Briefing Note 7.1—Stabilized Approach* (ALAR Tool Kit), *Flight Safety Digest* 19 (August–November 2000): 133–138.

20. Khatwa and Roelen, "An Analysis," 1–45.

21. Flight Safety Foundation Aviation Safety Network, *Accident Description* (Air Inter, A320, near Mt. Saint-Odile, France, January 20, 1992). Available online at Aviation-Safety.net/database/record.php?id=19920120-0.

22. J. Paries, "Investigation Probed Root Causes of CFIT Accident Involving a New-Generation Transport," *ICAO Journal* 49 (July/August 1994): 37–41.

23. Federal Aviation Administration, *Installation of Terrain Awareness and Warning System (TAWS) Approved for Part 23 Airplanes*, AC No. 23-18 (Washington, DC: FAA, June 14, 2000).

24. Tom Duke, "Conquering CFIT," *Air Line Pilot* 65 (March 1996): 10–14.

25. Federal Aviation Administration, *Installation*.

Abbreviations & Acronyms

AAIB	Air Accidents Investigation Branch (U.K.)
AC	advisory circular
AC	altocumulus
ACAS	airborne collision avoidance system
ACCAS	altocumulus castellanus
ACSL	altocumulus standing lenticular
ADM	aeronautical decision-making
ADS-B	automatic dependent surveillance-broadcast
AFCS	automatic flight control system
A/FD	Airport/Facility Directory
AFM	Aircraft Flight Manual
AFSS	Automated Flight Service Station
AGL	above ground level
AIM	Aeronautical Information Manual
AIRMET	airmen's meteorological information
AOA	angle of attack
AOPA	Aircraft Owners and Pilots Association
APU	auxiliary power unit
ARTCC	air route traffic control center
AS	altostratus
ASDE	airport surface detection equipment
ASDE-X	airport surface detection equipment (model X)
ASOS	automated surface observing system

ASRS	Aviation Safety Reporting System
ATC	air traffic control
ATD	aviation training device
ATIS	automatic terminal information service
ATP	airline transport pilot
ATSB	Australian Transport Safety Bureau
CAA	Civil Aeronautics Administration
CAB	Civil Aeronautics Board
CAF	Canadian Armed Forces
CAMI	Civil Aerospace Medical Institute (branch of FAA)
CARs	Canadian Aviation Regulations
CB	cumulonimbus
CC	cirrocumulus
CCSL	cirrocumulus standing lenticular
CDTI	cockpit display of traffic information
CDU	control display unit
CFI	Certificated Flight Instructor
CFIT	controlled flight into terrain
CFS	Canada Flight Supplement
CG	center of gravity
CI	cirrus
CIP	current icing product
CLR ICG	clear icing
CO	carbon monoxide
CRM	crew resource management
CS	cirrostratus
CSFF	cold-soaked fuel frost
CTAF	common traffic advisory frequency
CU	cumulus
CVR	cockpit voice recorder
CWS	control wheel steering
CYA	alert area (Canada)
DA	decision altitude
DCS	decompression sickness
DEP	design eye position (also called DERP)
DERP	design eye reference point (also called DEP)
DME	distance measuring equipment

DZ	drizzle
ECS	environmental control system
EFAS	En Route Flight Advisory Service
EFB	electronic flight bag
EGPWS	enhanced ground proximity warning system
EPT	effective performance time (formerly TUC)
ETA	expected time of arrival
ETOPS	extended-range twin-engine operations
FA	area forecast
FAA	Federal Aviation Administration
FAF	final approach fix
FARs	Federal Aviation Regulations
FCU	flight control unit
FD	flight director
FDR	flight data recorder
FE	flight engineer
FG	fog
FIC	Flight Information Centre (in Canada)
FIP	forecast icing product
FL	flight level
FMS	flight management system
FO	first officer
FPA	flight path angle
FPD	freezing-point depressant
fpm	feet per minute
FRST	frost
FSS	Flight Service Station (or Specialist)
FTD	flight training device
FZ	freezing
GA	general aviation
GAO	General Accounting Office
GPS	Global Positioning System
GBT	ground-based transceiver
GPWS	ground proximity warning system
GR	hail
GRC	Glenn Research Center
HAT	height above threshold

HEMS	helicopter emergency medical services
HOT	holdover time
hPa	hectopascal
IAS	indicated airspeed
IC	ice crystals
ICAO	International Civil Aviation Organization
ICGIP	icing in precipitation
ICTS	ice-contaminated tailplane stall
IFR	instrument flight rules
ILS	instrument landing system
IMC	instrument meteorological conditions
IOE	initial operating experience
IPS	ice protection system
KIAS	knots indicated airspeed
KTS	knots
LLWAS	low level wind shear alert system
LLWS	low level wind shear
LOC	localizer
LOC	loss of control
L/W	length-to-width ratio
MAC	midair collision
mb	millibar
MCTOW	maximum certificated takeoff weight
MDA/H	minimum descent altitude/height
MEA	minimum en route altitude
MEF	maximum elevation figure
METAR	routine weather report
MHz	megahertz
MOA	military operations area
MOCA	minimum obstruction clearance altitude
MORA	minimum off-route altitude
MSA	minimum safe altitude
MSAW	minimum safe altitude warning systems
MSL	mean sea level
MTR	military training route
MX ICG	mixed icing
NAS	national airspace system

NASA	National Aeronautics and Space Administration
NDB	non-directional beacon
NIOSH	National Institute for Occupational Safety and Health
NLR	National Aerospace Laboratory (Netherlands)
NM	nautical mile
NMAC	near MAC
NORDO	no radio
NS	nimbostratus
NTSB	National Transportation Safety Board
NVFR	night visual flight rules
OAT	outside air temperature
OE/D	operational (ATC) error/deviation
OROCA	off-route obstruction altitude
OTC	over-the-counter medication
PAPI	precision approach path indicator
PD	pilot deviation
PF	pilot flying
PIC	pilot in command
PIREP	pilot weather report
PL	ice pellets
PM	pilot monitoring
PNF	pilot not flying (old term for PM)
PWA	Pacific Western Airlines
PWS	predictive wind shear warning
QFE	station pressure
QNH	MSL pressure
RA	rain
RA	resolution advisory
RIME ICG	rime icing
RNAV	area navigation
RTO	rejected takeoff
RVR	runway visual range
RWS	reactive wind shear warning
SAFO	Safety Alert for Operators
SC	stratocumulus
SCSL	stratocumulus standing lenticular
SD	spatial disorientation

SG	snow grains
SIGMET	significant meteorological information
SLD	supercooled large water droplet
SLOP	strategic lateral offset procedures
SN	snow
SOPs	standard operating procedures
SPECI	special weather report (non-routine)
SRM	single-pilot resource management
ST	stratus
TA	traffic advisory
TAF	terminal aerodrome forecast
TAWS	terrain awareness and warning system
TC	Transport Canada
TCA	terminal control area (formerly Class B airspace)
TCAS	traffic-alert and collision avoidance system
TCU	towering cumulus
T/DP	temperature/dew point
TDWR	terminal Doppler weather radar
TIS-B	traffic information service-broadcast
TOGA	takeoff go-around
TS	thunderstorm
TSB	Transportation Safety Board of Canada
TUC	time of useful consciousness (old terminology for EPT)
UNICOM	universal communications
VASI	visual approach slope indicator
VFR	visual flight rules
VHF	very high frequency
VMC	visual meteorological conditions
VNAV	vertical navigation
VOR	VHF omnidirectional range
V/PD	vehicle/pedestrian deviation
V_{REF}	landing reference speed
V/S	vertical speed
WAAS	wide area augmentation system

Index

optimistic bias *89–91*

oxygen systems *132, 137–138*

Pacific Southwest Flight 182/Cessna 172 *25, 31, 45–46*

Pacific Western Flight 314, Cranbrook, BC *9, 15*

Pan Am Flight 1736 *1*

party-line information *41*

Payne Stewart, Learjet 35 *129*

personal weather minimums *94–95, 100, 101*

physiological deficient zone *130*

physiological factors
  and midair collision avoidance *32–35*
  and night flying *149–152*
  high altitudes and *128–134, 139–140*
  of spatial disorientation *187–190*
  and visual illusions *145*

physiological limitations of vision *32–35, 149–152*

pilot deviations (PD) *7*

pilot monitoring (PM) *77, 121, 209–212*

polished frost *68*

Portland, OR, United Airlines DC-8 *208*

precautionary landing. *See* off-airport landing

presbyopia *152*

pressure, self-induced and from others *92–93*

PRICE check *137*

primary ice detection system *65*

progressive taxi instructions *13, 153*

Quincy Airport, United Express 5925 *17–18*

radio communication *15, 17–18, 41–42*

rain on windscreen illusion *172*

rate of catch *58*

read-back/hear-back problem *207*

Reba McEntire, HS-125 *146*

relative brightness cue *168, 172*

relative size cue *35, 36, 164–168*

rime icing *59–60, 70*

roll upset *55, 63, 75*

Ron Brown, Croatia *200*

Roselawn, IN, American Eagle 4184 *55, 61, 63*

runway dimension illusions *164–167*

runway excursion *163*

runway incursions *1–18*

runway lighting illusions *168*

runway markings *10*

runways, crossing *3–4, 11, 13*

Sabreliner, Kaunakakai, Hawaii *169, 171*

San Diego, PSA Flight 182/Cessna 172 *25, 31, 45–46*

scuba diving *140*

scud running *86, 92, 96–97*

sea breeze *111*

see-and-avoid concept, limitations of *30–37, 40*

self-serving biases *90*

single-pilot resource management (SRM) *39*

situational awareness *202–203*

SkyWest Flight 5569 *7, 14, 144*

sloping runway illusions *161–164*

sloping terrain illusions *163–164*

Sochi, Russia, Armavia Airlines A320 *187*

somatogravic illusions *187–188*

somatosensory illusions *183, 187–189*

spatial disorientation *83–84, 181–193*
  classifying *190*

special use airspace *46*

stabilized approaches *119, 201, 211*

stagnant hypoxia *134*

stall warning bias *76*

standard operating procedures (SOP) *14, 209, 215*

stepped-on transmissions *9, 15*

sterile cockpit rule *16, 45, 77, 209, 212*

strategic lateral offset procedures (SLOP) *47–48*

St. Thomas, Virgin Is., Cessna 402 *169, 171*

sudden decompression *132–134, 139*

supercooled large water droplets (SLD) *61–63, 70–73*

supplemental oxygen *128–129, 135*

surface event *2*